THE WEATHER

THE WEATHER

FIREFLY BOOKS

contents

foreword

by Donal MacIntyre

Left: Sunlight is refracted through atmospheric ice crystals to produce a sundog over Alaska. In northern climates the sundog is also known as a 'winter rainbow'.

Over 1000 miles (1600 kilometers) inside the Arctic Circle, sitting precariously on the back of an overladen sleigh pulled by thirteen specially bred huskies and driven by two elite Danish soldiers, I caught a special glimpse of how the world and its inhabitants adapt to the most extreme of environments.

It was -40°C (-40°F) on the eastern coast of Greenland at the Mestervik army base in the middle of nowhere, accessible only by satellite phone and the Danish mail service. It was 10 p.m. and a cold breeze was cutting across at 30 mph (48 kph). A light dusting of snow was kicked up and swirled as the dogs (part-wolf and part-Inuit husky) drove off their hind legs, howling and barking with the thrill of an expedition. The expedition in question was a short one—in fact, a short hop and parcel dispatch. The parcel was me, and the repository was a snow hole, on the side of a hill, that was to be my home for the night.

The night was glorious with moonlight throwing a pink hue on a glossy white landscape. Earlier in the evening the northern lights had danced, sea green, across the sky for us; taking on the perfect shape of a sea horse. I watched mesmerized, sinking and slipping on the snow until I lay flat on my back for ten minutes dazzled by nature's greatest show. The wispy green apparition faded peaceably away. Still drunk on the experience, and a couple of glasses of red wine with Commander Norritt and his Danish SAS unit, I prepared for my night in the snow hole—an experiment constructed by the producer to make my life as uncomfortable as possible. One of these days they'll kill me off.

The next morning as we left the compound at quite a pelt I noticed one of the dogs in the middle of the pack veer left off the track. His body turned but the momentum of the pack kept him in line. The dogs with their thick fur might be oblivious to the cold, but dog number seven was not oblivious to the Arctic hare sitting tall about 45 yards (50 meters) off the moonlit track.

Soon the others caught the scent and veered off in the direction of the snow-white hare—in the pink glow its outline was discernible only by its dark eyes. As thirteen howling wolf-dogs raged towards it, the hare moved not a jot. It knew that, more than any creature there that night, it was adapted perfectly to this environment; so instead it just stood tall and surveyed its patch, showing no sign of fear.

The dogs jangled in their chains, a triumph of breeding and adaptation, struggling against the army handlers who crackled them into line. The men, wrapped in down, were trained to live unsupported with the dogs, in the most extreme conditions on the planet, for three months at a

Above: Huskies are an impressive example of successful breeding and adaptation for such extremes of climate as those experienced in the Arctic.

time. The army's elite force learned to tolerate (just) their surroundings and each man had his own bag of tricks—and socks—to deal with what is frankly an inhuman climate. And then there was me. A freezing, terrified Londoner—trained for nothing, and adapted to no climate except one where central heating and a good pub are nearby.

But what a world the weather offers us! And how remarkable it is that we can adapt and mold and mutate to survive in the most arduous of climatic and environmental conditions. I've traveled the globe tasting and touching the worst and the best of the world's weather—from the Sahara to the tropics, from the blast of the Fastnet wind to the silence of the doldrums and from the heat of Arizona to the wet of the Belizian jungle—and the truth is that no animal on the planet can survive its variety and extremes of weather and climate better than man. In the Arctic the hare is entitled to disagree, but that is its kingdom. Across the globe, the awesome and inexplicable truth is that the human body is a remarkable machine that has habituated itself to a wild and turbulent world. I use "man" in the generic sense. Me—I'm still lost at sea without my central heating and my pub.

DONAL MACINTYRE, 2002

wea

Weather is in us and all around us, a huge interconnected force of chaos, creation and destruction. It is also the thin veil that shrouds our planet and defines the limits within which we can stay alive: the temperatures below and above which our bodies will cease to function, the forces of wind and flood that we can withstand, or the electrical charge that can instantly destroy us.

chapter one

therworld

seeds of weather

Sit on a hillside overlooking the sea on a warm summer's day, relax in the golden heat of the afternoon, pluck the head from a dandelion and blow away the seeds—watch them scatter in the wind and you will have become part of the weather. As the tiny seeds fly through the air at your behest, the microturbulence you have created will meet updrafts of warm air from the field and little eddies of wind will spiral upwards into the sky above. More air will be drawn into the void left behind by the rising air, creating a gentle wind that plays on your face, the beginnings of a shore breeze. Meanwhile, far above you, some of the tiny particles of pollen from the seeds will continue to ascend, reaching so high into the air that they find themselves in a region of icy cold. Here they may form a "seed" of another kind—the heart of an ice crystal—as even tinier molecules of water cluster around them and hold fast. In turn these microscopic structures of ice will play in the up- and downdrafts in the weird world at the top of what is now a spreading thundercloud, growing to become pellets of hail or falling and inexorably melting out as rain that eventually drenches the grass around you. Instead, as the cloud spreads out in the familiar anvil shape of a summer storm, some of the crystals and their passengers of pollen may be carried far across the globe by the superfast winds of the jet stream, eventually perhaps sprinkling down as snow on the cap of a distant mountain and there melting, to become a torrent that cuts its way down a river gorge, across a parched flood-plain and out towards the sea. And the water that then flows from this estuary on the far side of the world may in turn undergo an epic journey through the depths of the ocean, to return perhaps five hundred or a thousand years later to the very spot where you are sitting. It may drift in once again as a cloud, to rain on the ground and to be drawn up by the roots of the ten-thousandth generation of the dandelion plant that you held in your hands. For the weather is something that is in us and all around us, a huge interconnected force of chaos: a force of creation and destruction, a thin veil that shrouds our planet and allows us to live and thrive.

wind, wet, hot and cold

We live in the "Goldilocks zone"—neither too hot nor too cold, too wet nor too dry, too windy nor too calm. By the sheer chance of planetary mechanics, Earth is the perfect home. Our neighbors in the solar system, Venus and Mars, are in just the wrong position for the weather to have worked out perfectly. Venus is too close to the sun and its atmosphere is a searing, boiling greenhouse of immense pressure that would simultaneously fry you, crush you and dissolve you in acid were you to venture out on to its surface, while Mars is too far away and is a frozen world with almost no atmosphere at all. But on Earth the forces of geology, biology and chemistry flow hand in hand, behaving as a single living entity, and the weather we experience is effectively the breathing of that living planet.

For our life is intricately linked to the weather. We depend on its remarkable system for the air we breathe, the water we drink and the growth of the food we eat. The weather defines the limits within which we can stay alive: the temperatures below and above which our bodies will cease to function, the forces of wind and flood that we can withstand, or the electrical charge that can instantly destroy us. Throughout human history, the weather has been a dominant factor in our lives. Drought and rain have always been the controllers of famine and plenty, storms and calm have limited our global exploration and ocean trade, while individual lives have been saved by a shower of rain or swept away by a gale. Civilizations have fallen and wars have been won and lost on the turn of a few days' weather. It dictates the cultural attitude of whole nations, it has been at the center of religious belief for millennia, it makes us happy, it makes us gloomy, it reassures us and surprises us, and its unpredictability tests us to the limit. The weather is our friend and our enemy in equal measure—it is at the heart of our existence.

Above: The dry, freezing cold surface of Mars with its empty, pink sky, photographed by the Pathfinder lander in 1997.

In a sense all of our weather began a million years ago, and 93 million miles (150 million kilometers) away inside the sun, which is in fact a massive star. The heart of the star is an almost inconceivably huge thermonuclear fusion reactor—a hydrogen bomb which reaches a temperature at its core of 57,000,000°F (14,000,000°C)—but the energy generated there goes through a series of epic permutations before it reaches the surface of the Earth. It begins as gamma rays produced by the nuclear reaction, but these quickly degrade to X-rays. These radiate from the core until the temperature has cooled enough to allow volumes of gas to carry them rapidly in giant cycles of convection to the surface. There the temperature is much more comfortable—a mere 11,350°F (6,300°C). This journey of energy from the core to the sun's surface can take a million years or more. As they near their journey's end the X-rays lose energy, so that what emerges is first ultraviolet, and then finally the bright visible light that we see at the surface. There, in the turmoil that exists, the constant leaping of electrons to and from different states of energy, as they orbit the nuclei of their atoms, releases photons or light energy. These photons radiate out towards the planets, traveling at the astonishing speed of 186,000 miles (300,000 kilometers) a second. In just over eight minutes the photons strike the Earth where, as each one is absorbed by the ocean, by a cloud droplet, by a rock, a plant, an animal, or you or me, it adds to the energy that warms the surface of the planet and adds to the force that is at the heart of our weather.

Right: The thin layers of the Earth's visible atmosphere—the veil of weather on which we all depend.

weather machine

The radiation from the sun heats the surface of the Earth, which in turn heats the air above it, causing it to rise. Air above a warm part of the planet will rise faster than the surrounding cooler air, so more air slides in to take its place in the low-pressure void left behind. So wind is born. And that is the heart of the weather machine: heat and pressure differences bring cyclones, depressions, thunderstorms and hurricanes, all blown across the surface of the globe by the wind.

But if wind is the engine of the weather, water is the fuel. Our weather consists of a constant interplay of water as vapor, liquid and solid ice. As we will see in later chapters describing the formation of rain, snow and storms, so much depends on water molecules changing from one state to the other that it is not surprising to find that our planet lies at precisely the right distance from the sun for water to exist in all three states at the same time. Alone among the planets of the solar system, Earth has average temperature and pressure that are very close to what scientists call the "triple point" of water—the combination of temperature and pressure at which water can exist in all three of its physical states. This property of our planet is one which allows us to exist.

Warm air, dry air, cold air and wet air all behave in different ways, and it is those qualities of air that make it move and drive the engine of the weather. Although it may not seem obvious, moist air is less dense than dry air, and so will always tend to rise above it. This is because of a very simple piece of physics. A water molecule is made up of one oxygen atom and two hydrogen atoms (H_2O). Because hydrogen atoms are the smallest and lightest atoms that exist, a molecule of H_2O is much lighter than a molecule of either oxygen (O_2) or nitrogen (N_2), which are what dry

Above: The Earth from the Apollo11 spacecraft. The blue refraction of sunlight is absorbed by the atmosphere to leave

the pink color of sunset.

air consists of. In moist air, water-vapor molecules have replaced some of the oxygen or nitrogen molecules, one for one, so any given volume of moist air contains more lightweight molecules than does the same volume of dry air, and therefore it is lighter and less dense. Warm air is less dense than cold air because, when they are warm, all air molecules are more active, move around more and take up more space—the air expands. Thus a given volume of warm air will simply contain fewer molecules than that of cold air, so it will be less dense.

In a nutshell, if there is warm, moist air it is going to move up and, at its very simplest, all of our various forms of weather have that single action at their heart: the air rises, in comes the wind below; as the air rises further it cools, the water vapor condenses to cloud and down comes the rain. That's all there is to it.

clouds

The most direct and obvious reminder of moving air's simple ability to create the weather is clouds. Clouds consist of moist air that has risen, cooled and released tiny droplets of water that are light enough to remain airborne. As we shall see in later chapters, clouds bring rain, snow, thunder and lightning; they are made by wind, and they help to make wind; they bring us the rainbow. For almost all centuries of human civilization they have been thought of simply as "clouds," with descriptions steeped in country weather lore (like "mackerel sky" and "mares' tails"), and it was only at the beginning of the nineteenth century that an attempt was made to classify them precisely into the different categories and types that we know today, in order to understand and explain them better. In the climate of a strong desire to classify and order nature that had emerged from eighteenth-century thinkers like Linnaeus, an English naturalist called Luke Howard came up with four Latin names in 1803 with which to define the different cloud types: *stratus* ("spread") for sheets of cloud; *cumulus* ("heap") for fluffy cloud; *cirrus* ("curl") for wispy cloud; and *nimbus* ("rain") for a rain cloud. These four words are used today in various combinations to define our clouds and order them into the ten different types that are agreed on the world over—ten types that to the practiced eye still provide the most consistent signals of what the weather has in store for us at any one time.

Above: A layer of stratus cloud covers the sky at high altitude.
Right: Patches of stratocumulus cloud at the end of a day are the last remnants of a collapsed afternoon thunderstorm.

TYPES OF CLOUD

Clouds are classified according to the height of the cloud base above the Earth: high, middle or low, with a fourth group for clouds that are changing height and growing vertically.

All the high-level clouds are white in color, because at these extreme altitudes they are made up only of ice crystals. *Cirrus* is thin, wispy streaks of white, splayed out by the high-level winds and often known as "mares' tails." Its appearance can often be the first warning of the arrival of a weather front. *Cirrostratus* is a sheet of cloud that may cover the entire sky but is so thin that the sun or even the moon can be clearly seen through it. Often cirrostratus creates a halo around the sun or moon. Thickening cirrostratus clouds are a strong predictor of forthcoming rain or snow. *Cirrocumulus*, more rarely seen, is small, round puffs of white, often appearing in long rows of ripples across the sky, which can catch the light of a setting sun to beautiful effect.

At the middle level, the air is warm enough for water droplets to be the principal constituent of the clouds. *Altocumulus* forms as a white or gray, fluffy mass, spread out in waves across the sky. It is a sign on a summer morning that thunderclouds may build up by the afternoon. *Altostratus* is the cloud that gives us a "watery sun"—vast even, gray sheets that remove all shadows from the surface of the Earth. Its slowly lowering height and evenness over huge areas at a time are signs of prolonged rain to come.

Closer to our heads are the gray, wet-looking clouds that are so often the deliverers of the rain itself. *Nimbostratus* is that solid layer, with darker wisps of ragged cloud a little lower than the main

FOUR TYPES OF CLOUD
[1] Cumulus: with fluffy cotton-wool appearance, these clouds are associated with fair weather.
[2] Cirrus: blown by the high altitude jet stream winds, to form characteristic "mare's tails".
[3] Cirrocumulus: very high altitude clouds, creating a "mackerel sky".

mass. It wafts across on the wind and brings almost continuous light rain or snow. *Stratocumulus* looks like patches of lumpy cloud, ranging in color from light to dark gray, with glimpses of longed-for blue sky in between the cracks; these often appear towards the end of the day, because they frequently mark the remnants of a much larger cumulus cloud that is in decline. *Stratus* is the most depressing cloud of all. An even, flat, gray cloud that fills the sky, often hanging low in valleys and masking the tops of hills, it does not bring rain as such, but instead delivers a constant dampness or drizzle.

Finally there are the powerful clouds that really impress us, those that can be seen building to greater height, with fluffy cotton-wool tops. *Cumulus* clouds usually have flat bases, usually gray with flecks of white, with a rounded dome of white above. Although many of them can fill the sky, each is clearly an individual, with blue sky around it. Look closely and you can see the swirling edges at the top of the cloud: that ever-changing boundary, marking the upper limit of the rising moist air, is the front line of a constant battle of physics as its molecules of water vapor rise, cool and condense to form cloud droplets.

Cumulus clouds that do not rise far are known as "fair-weather cumulus," and are often present on warm summer mornings. But by afternoon they can build up like a growing cauliflower to produce "towering cumulus." from which showers may fall. If such a cloud continues to grow taller, it becomes the *cumulonimbus* cloud of a thunderstorm. These giant individual clouds rise to the limit of the lower atmosphere, where they can be seen spreading out into the familiar anvil shape as they are blown by the high-speed winds that race around the planet far aloft. These clouds take on remarkable properties of their own (See Chapter Three), and become truly fearsome sources of rain, thunder, hail and lightning.

[4] Cumulonimbus: the towering head of a thundercloud, blown out to an "anvil" shape by high altitude winds.

Left: The names and forms of different kinds of cumulus cloud depend on their height, and in the case of cumulonimbus, on their growth.

exploring the sky

Finding out what went on in the world of the clouds was the purpose of the extraordinarily foolhardy adventures of some nineteenth-century aerial explorers. Yes, the clouds make lightning (thanks to Benjamin Franklin for all that), they make rain and snow (even the Ancient Greeks had worked that out), and they consist of droplets of water (the French philosopher René Descartes had got that one cracked), but what were they actually like to be in? After the Montgolfier brothers made their famous ascent over Paris in a hot-air balloon in 1783, a succession of intrepid flyers went up to have a look. Many carried rudimentary instruments with them to measure the atmosphere through which they flew (crude versions of the barometer and thermometer had been invented some 150 years previously) and gradually a map of the layer of air above our heads was built up. Falling temperatures and lower pressures were experienced the further up the explorers ventured, and many reported strange ill-effects, like bleeding from the nose or difficulty in breathing. Some fared far worse.

In 1862 James Glaisher, a founder of the British Meteorological Society, was dispatched to investigate precisely what occurred at height, along with an experienced balloonist, Henry Coxwell. The balloon was a gas-filled canopy, with sandbags for ballast, and carried a variety of instruments to measure air temperature, pressure and humidity, magnetic fields and the spectrum of the sun. The two men even had a crate of pigeons to test the effects of altitude on animals other than themselves.

The morning of the flight dawned with a threatening sky, but the men wanted to go ahead, so the balloon was launched right into the clutches of a rising rainstorm and they were sucked up into a thundercloud. They suffered the terror of rising through the thunder and lightning of the storm only to emerge miraculously out of the side of it. They continued to climb, taking measurements as they went. At 16,500 feet (5,000 meters) and 18°F (–8°C) a pigeon was tossed over the side and fluttered chaotically towards the earth. At 23,000 feet (7,000 meters) another luckless pigeon plummeted far from view before regaining some kind of aeronautical control. At 26,000 feet (8,000 meters) the last two pigeons refused to get out of their cage. Then disaster: concentrating on their experiments, Glaisher and Coxwell failed to notice that the release-cord for the gas valve was tangled up out of reach. They had risen to 30,000 feet (9,000 meters) and with no way of letting the gas out of the balloon they seemed doomed to go on rising, becoming ever colder and ever more short of oxygen.

By now breathing and movement were painful, but Coxwell managed to climb into the rigging to try to reach the cord, then found that his fingers literally froze to the metal structure that held the gondola and that he was trapped, unable to move up or down. Glaisher meanwhile was losing his sight and his mental faculties as he struggled for oxygen, and failed to understand when Coxwell yelled for help. But Coxwell managed to keep control and caught the swinging valve-cord in his teeth. Twice he tugged at it with his head, once breaking a tooth and once tearing through the flesh of his cheek, and then for the third time he bit on to it, clenching with all his might, and jumped. Tearing his frostbitten fingers from the rigging, he crashed with good fortune back into the gondola, the gas valve wrenched open above them and their gradual descent, and recovery, began.

They had reached 36,000 feet (11,000 meters), close to the limit of human survival, and

Above: The departure of the Montgolfier brothers at a balloon rally in France.

they had charted their way to the top of the weather. There were more balloonists, more courage, more disasters, until the limits of balloon flight were reached in the 1950s, touching altitudes of 102,000 feet (31 kilometers) where it is possible to see thin layers of the atmosphere that wrap, like the skin of an onion, around the planet. Earth's atmosphere stretches 80 miles (130 kilometers) from the surface of the planet to reach at its furthest point the ionosphere, the zone in which the spectacular northern and southern lights of the aurorae play. Every layer has a vital role in the safeguarding of life on the planet but the weather as such exists only up to little more than the height of our tallest mountain. In the layer known as the troposphere.

At its thickest, over the Equator, the troposphere is only 10 miles (16 kilometers) high, falling to just 4 miles (7 kilometers) over the poles. And above most of the populated areas of the northern hemisphere only a 6-mile (10-kilometer) thickness of weather lies above us. On a planetary scale this thin veil is an almost non-existent place in which we survive, a thin blue gas which sustains all life. Fortunately it is enough: at any one time we are wrapped in about 5,600 trillion tonnes of air; each one of us breathes about ten million times a year, using only about 5 tonnes in that time—so there is plenty of it to go around. How that air moves and behaves, along with the myriad particles of dust and molecules of water that it carries with it—in other words how it creates weather—is governed by the forces of heating and cooling described earlier (pp16–17), but also by another property of our Earth.

spin of the earth

The way the planet itself moves is ultimately one of the keys to the weather. The Earth is spinning as it orbits the sun: once every twenty-four hours it rotates, creating our days and nights, allowing the sun to warm us and the darkness to let us cool. As will be explained in the chapters that describe the wind, the wet, the hot and the cold that make up our weather, it is this heating and cooling that lie at its heart. But there is another factor that the movement of the planet brings into play—movement itself.

As the planet spins, so the atmosphere that shrouds it and the oceans that sit on its surface spin too, and the effect is to create a precise pattern of motion that governs every wind, every cloud and every ocean current on the globe. Worked out by a French physicist, Gustave-Gaspard Coriolis in 1835, this has become known as the Coriolis effect, and its net result is that air flowing in the northern hemisphere will tend to turn to the right, and air flowing in the southern hemisphere will tend to turn to the left. It is this rotational property of the airflow across the surface of the planet that gives the weather its sense of direction. That, too, has ever been one of the keys to forecasting what will happen next.

Above: Gustave-Gaspard Coriolis (1792–1843)

CORIOLIS EFFECT

The spinning Earth has considerable rotational velocity, and this is greatest at the Equator, where the circumference of the planet is largest. In fact, at the Equator the surface of the planet is spinning at 1,036 miles (1,667 kilometers) an hour, and this rotational speed gets less and less the further you are from the Equator, until at the poles it falls to nothing. So anything not directly connected to the surface of the planet, be it a molecule of air in the wind, water in an ocean current, or even a ball thrown through the air—anything which moves away from the Equator towards the pole—will move eastwards through space much faster than the point at which it was aiming when it started. So by the time it reaches a higher latitude its rotational speed will have moved it further to the east relative to a point on the planet's surface at that latitude, which is rotating more slowly. Its movement will appear to have been deflected to the east. In the northern hemisphere that would mean to the right, in the southern hemisphere to the left. Equally, anything traveling from the poles would lag behind the faster-moving surface of the planet nearer the Equator. So it would be deflected to the west—again to the right in the northern hemisphere and to the left in the south. The net result of this simple fact of physics is what gives direction to cyclones, anticyclones, hurricanes, circulating ocean currents (and water down drains, if no other factors got in the way). And as we shall see, that spin is what brings the weather to reach its most terrifying forms.

If you find this all too much to believe, the story of German gunners in the First World War may convince you. The kaiser's artillery divisions built two huge guns towards the end of the war, precisely so that they could bring terror to the unsuspecting inhabitants of Paris. The guns, called Big Bertha and Long Max, were capable of firing a 210-millimeter howitzer shell over a distance of more than 60 miles (100 kilometers), from the nearest German front line to the French capital. The gunners had perfect coordinates and the shells were well able to reach their targets. But they consistently kept missing, falling about a ½ mile (kilometer) west of the city. It is thought that the artillerymen had fallen foul of the Coriolis effect!

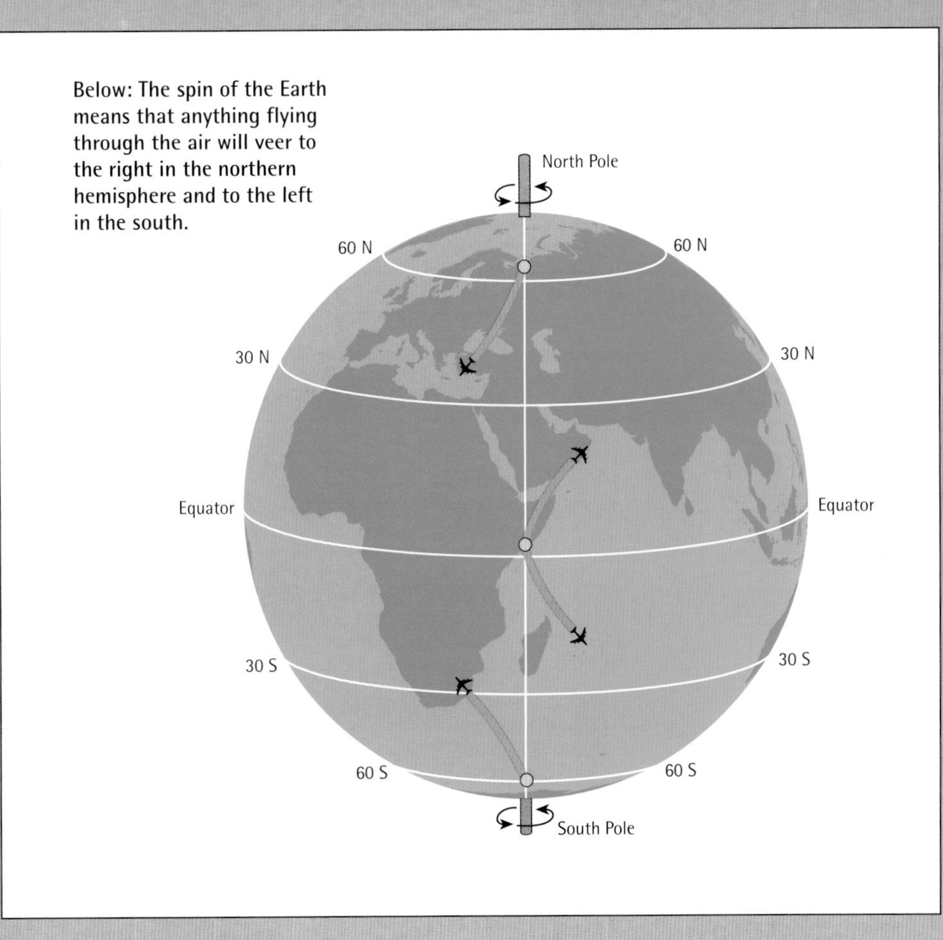

Below: The spin of the Earth means that anything flying through the air will veer to the right in the northern hemisphere and to the left in the south.

Above: A "red sky at at night," suggesting dry weather approaching from the west.

red sky at night

Red sky at night, shepherd's delight.
Red sky in the morning, shepherd's warning.

We all know this rhyme, and we all feel cheered at the sight of a strong pink sunset. It is probably the most famous of all the ancient weather lore that has been passed down from generations of sky-readers before us. And, like so much of that traditional knowledge, it is well worth heeding, for this simple adage of weather forecasting has more than a grain of truth to it. The color of an evening sky is due entirely to the nature of the air through which the dying sunlight passes before it reaches our eyes, because when the sun is low on the horizon its rays have to travel through a great thickness of atmosphere to get to us. If that air is dry, the dusty particles that will be floating in it will scatter and absorb sunlight, making it appear red to our eyes. So somewhere out there in the direction of the sunset is dry air. But why does that mean tomorrow will be fine? Well, the sun sets in the west and, because of the spin of the Earth, it happens that the prevailing winds in the mid-latitudes (where most of the populations that have inherited this age-old saying live) also come from the west. So the red sky at sunset will generally mean that drier air is on its way from the west; it probably marks that the weather is beginning to clear after a cold rain front has passed us by. In the morning a red sky to the east (where the sun rises) means the fine weather is probably moving away, so another front is likely to arrive soon. It may have taken meteorologists and physicists till the last century or so to work out the explanation, but the shepherds had it right all along.

"Red sky at night" has been handed down to us through many generations, and it even features in the New Testament. But three hundred years before the gospels were written one of Aristotle's pupils, a Greek philosopher called Theophrastus of Eresos, compiled a list of some two hundred "signs" that could be used to predict the onset of fair weather and foul. In many ways this was the first attempt to provide good empirical evidence on which to base a forecast of the weather, rather than simply putting what happened down to the random whim of the gods. Indeed, Aristotle himself had written a treatise on the weather, *Meteorologica* (from which the word "meteorology" comes), which attempted a scientific explanation of what happened in the sky, and had correctly worked out that the sun's heat caused water to evaporate into the air, and form cloud and eventually rain.

Many of the signs that Theophrastus noted are, like red sky, effectively the same as those passed on to us by succeeding generations of country folk, although few of those people were aware of the existence of any Greek weather lore. Today we know that there is a scientific basis for many of these sayings. For instance, the rhyme "Ring around the moon or sun,/Rain before the day is done" is based on the fact that a ring is formed due to light being refracted by the ice crystals in the high clouds that form ahead of an approaching warm front. The saying "Swallow high,/ Going to be dry" has a lot of truth in it because the insects that swallows feed on rise higher in the air with rising barometric pressure).

But of course many old country sayings are hopelessly wrong: for example, cows do not lie down when it is going to rain! But however rational the proverb, and however clearly based on scientific evidence it can now be seen to be, nothing will ever predict the weather 100 per cent. As Chapter Three will show, at the heart of the world's weather is a phenomenon which scientists have now learned formally to recognize: chaos, and predictable unpredictability.

satellites in space

The logic of the ancient Greeks—combined with the practical experience of a hundred generations of country folk—sufficed for our understanding of the weather until the scientific revolution of the seventeenth century when, as we shall see in later chapters, great names of science, such as Galileo, Descartes and later Benjamin Franklin, all played their part in laying bare the secrets of nature. But it is perhaps not surprising that precise explanations of the weather, and the more accurate predictions we are used to today, are only recent additions to our lives. For the weather operates on a global scale, and it is only in the last half-century that we have been able to see it directly in that way—to see what the weather really looks like. That milestone was reached as recently as April 1, 1960.

At 6.40 on the morning of that day a Thor-Able rocket was launched from Cape Canaveral in Florida, carrying a tiny satellite into orbit. It measured just 3 feet (1 meter) wide, and 18 inches (half a meter) high, and it lasted for a mere seventy-eight days in space, but TIROS (Television Infra-red Observation Satellite), as it was called, changed the world of meteorology forever. With a miniature television camera on board, TIROS sent back images that had only been imagined before. From its vantage point of 450 miles (725 kilometers) above us, it orbited the Earth every ninety-eight minutes, capturing the patterns of the clouds from one side of the planet to the other, from as far apart as the Pacific Ocean to central Europe. Its first fuzzy black-and-white image transmitted live from space revealed thick cloud over the Atlantic Ocean and a sunny day in Nova Scotia. A week later it spotted a typhoon in the Pacific, east of Australia. It was with TIROS-1, and the nine other TIROS satellites that followed, that we began to appreciate the intricacies of the weather as part of one huge system, constantly whirling above our heads. Within five years scientists had put together 450 TIROS images to create the first global image of the weather.

Above: A satellite image showing a tropical storm, a precursor to a hurricane, building up over the Atlantic Ocean.

Today we take for granted the extraordinary clarity of real-time animated images of hurricanes, storm fronts and cloud layers that arrive regularly from the array of satellites that sit in geostationary orbit around the planet, hovering above our heads at a tenth of the distance to the moon. We can see the weather twenty-four hours a day, anywhere in the world. Forecasters can watch weather systems emerging, evolving and dying before their eyes, enabling the science of weather prediction to reach an astonishing level of accuracy. But it all began with TIROS.

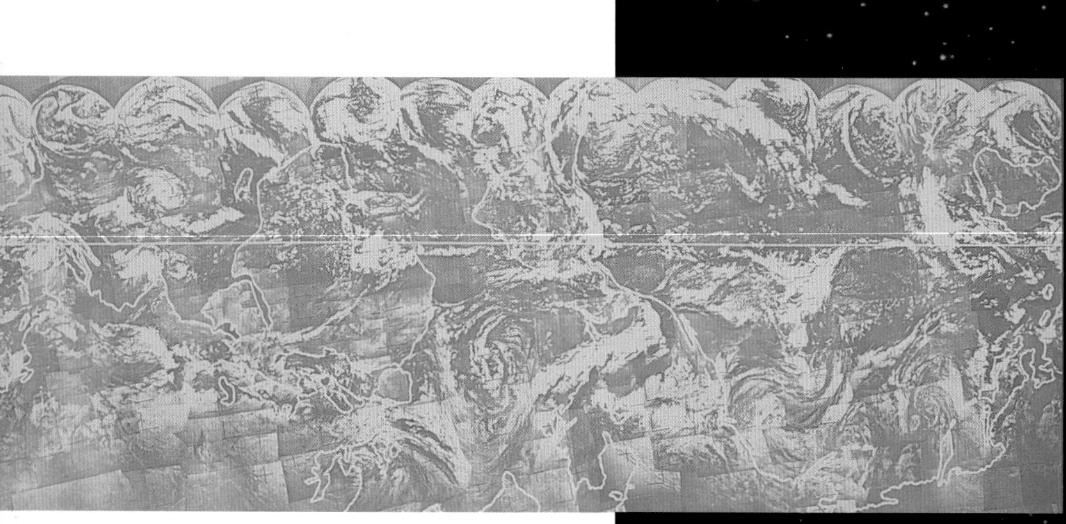

Above: The first complete view of the Earth's weather, a composite of images from TIROS on February 13, 1965.

Right: A computer-generated image of TIROS as it would have looked during the first-ever orbit of a weather satellite.

RAINBOW

A rainbow is probably the most beautiful of all weather phenomena. It is the one that has inspired poets and songwriters throughout the ages, and it is the one which has perhaps more religious and superstitious associations than any other in the many cultures of the world. By some, the rainbow was thought to be a bridge between this world and the spirit world, a reminder of God's covenant with Noah that he would not send another flood. Another popular explanation was that it was a pathway to wealth and good fortune, with a crock of gold at its end. In fact it is a simple optical effect that was first worked out by a Dominican monk, Theodoric of Freiburg, somewhere between 1301 and 1310. Using a spherical flask of water, a hexagonal crystal and parchment pierced with a pinhole, Theodoric came to the conclusion that a rainbow was created by light that was refracted as it entered a droplet of water and was reflected around the droplet's inner surface back towards the observer, getting refracted again on its way out. He was right.

A rainbow can be seen only when the sun is behind you as you look towards rain in the distance (or water from a garden hose!). Light from the sun enters each raindrop and, because the droplet has a different density from the air around it, the light is bent, or "refracted." Most of the light passes on through the droplet, never to return, so to speak, but, once inside the droplet, if any of the light rays strike the surface at the other side of the droplet at an angle greater than 48 degrees, they are reflected back, and then will be refracted again on their way out, heading back towards the viewer. Because the surface of the droplet is curved, each light ray will be refracted, strike the surface, reflect and be refracted again at a very slightly different angle, so that the light is spread out into its different constituent colors: red, orange, yellow, green, blue, indigo and violet. Light emerging at an angle of 42 degrees is red and at 40 degrees is violet, with the other colors in between. We see a single color from a single drop, so it takes a multitude of them to create the beautiful array of the full rainbow, with the "primary" arc appearing with red on the outside and violet on the inside of the curve. If the sun is close to the horizon, the rainbow will form a semicircle in the sky, if it is higher up the curve will be flatter; and if the angle of the sun is greater than 42 degrees, no rainbow is visible at all. A secondary rainbow can sometimes be seen faintly above the primary one, with the color sequence reversed. This is due to light getting a double reflection inside the droplet before re-emerging. It is even possible to get up to four fainter subsidiary arcs inside the primary one and three inside the secondary. But that would be quite a lucky sight.

Above: Droplets of water refract sunlight that reflects off them, to create our view of a rainbow.

Just as a red sky can portend rain or shine, so a rainbow, statistically, will also hint at the weather to come. A rainbow seen in the evening, with the sunlight coming from the west, will most likely foretell good weather because the wind is probably blowing from the west as well, carrying the rain away from you. But however accurate the rainbow forecast may be—and it isn't very—one of the most charming aspects of a rainbow is that each viewer sees the light from a very slightly different angle, refracted through different droplets, because of where they are relative to the sun.

So although a rainbow may seem the same to everyone watching, in fact each one is different—and everyone has their own personal rainbow.

the big picture

On a planetary scale the world's weather system is driven by two great global cycles: wind and water. Chapter Two will show how the winds that howl across the surface of the Earth are part of a slow movement of air from the warm Equator towards the colder poles and back again. No individual parcel of air can be traced on that journey, as it will be caught up in a constant, turbulent whirl of easterlies, westerlies, trade winds, cyclones and hurricanes. But the inexorable need for the hot air to get cooler and the cold air to get warmer, to even out in the atmosphere, is what drives the wind. In Chapter Three we will enter the water cycle and follow a droplet of water as it travels from ocean to air and back many times, in an extraordinary epic journey around the globe—a journey that can last for a thousand years.

There are two other great forces that govern the weather: hot and cold. As the Earth tilts slowly back and forth each year on its journey round the sun, so the cycles of hot and cold come and go. As the northern summer sun warms the deserts of Africa, dry winds spread north across Europe, hurricanes are born off the Atlantic coast and clouds of dust are carried across the ocean to become Caribbean sand. But as the Earth moves on and the sunshine shifts to the south, so summer crosses the Equator, and the cold waves of ice and snow surge south from the Arctic, bringing with them the cold that is our most enduring climatic enemy.

So the central four chapters of this book take an intimate look at the weather's four faces: wind, wet, cold and hot. But, as we are all now aware, our weather is changing; the climate is getting warmer and the weather is getting wilder—what will be the consequences of these changes for us? Chapter Six looks at how humans have learned to change the weather deliberately—and also inadvertently—and at the surprises that are in store for our future.

Above: True-color satellite image of the entire Earth's surface, showing tropical vegetation as green, arid regions as yellow and brown, the sea as blue and clouds and snow as white.

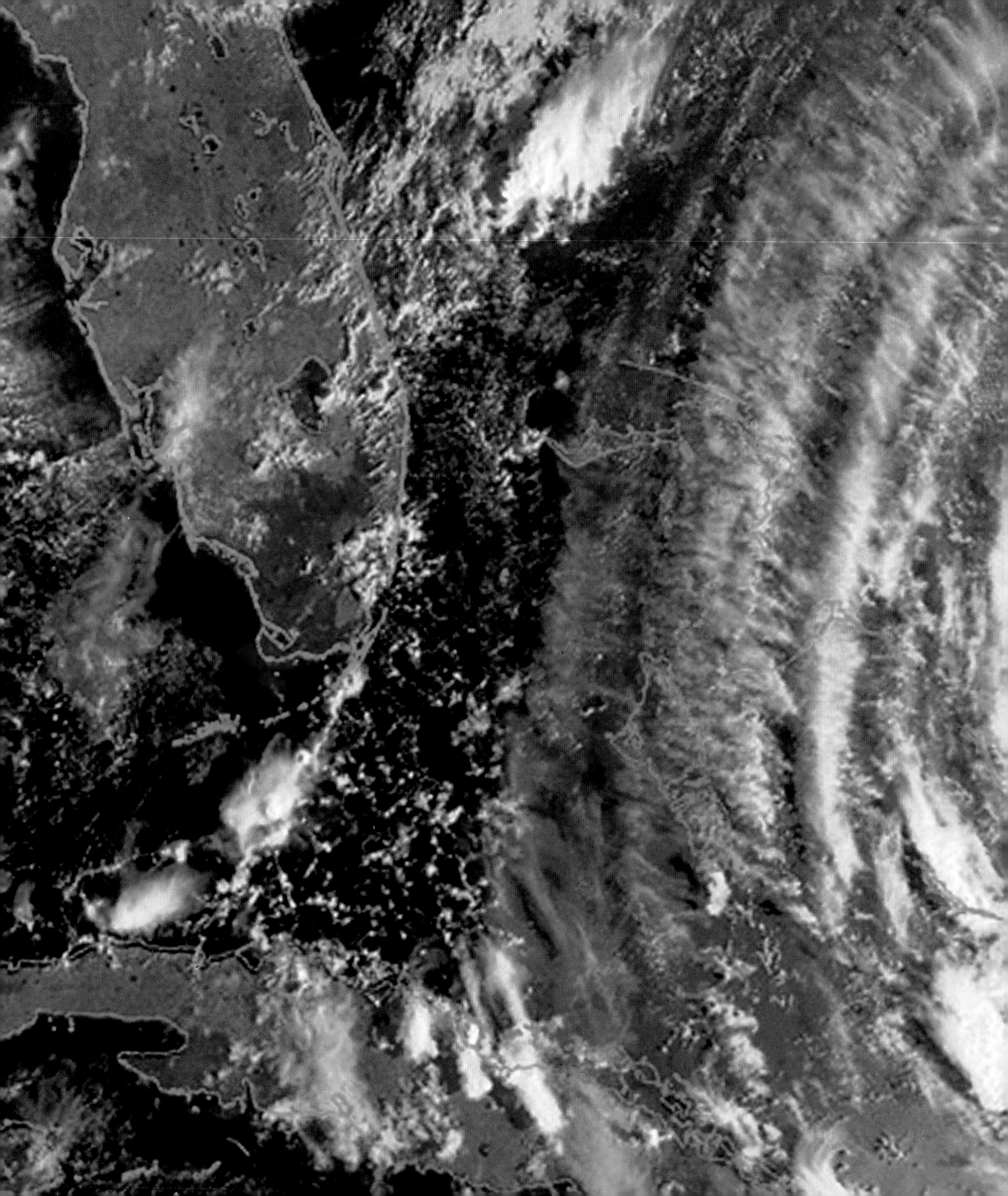

The wind carries the air we breathe as far as the North Pole and back, out to the edge of space, and above the highest thunderclouds. Sometimes air whips up waves in a surface breeze, sometimes it moves on to whirl in a tropical cyclone, sometimes it drives an icy snowstorm. Wind in a sense is the engine of the weather, and its story is one of adventure, tragedy and surprise.

chapter two

wind

deadly calm

It begins with nothing.

"After a while the sea was so still that scarcely a block creaked in all the maze of our tremendous rigging and the sails hung lifeless, with not even a gentle heave of the ship to send them slatting against the masts. There was not the slightest movement in the oppressive air, and the surface of the sea lay panting quietly, like a great heat-tormented beast. Day after day, we drifted in calm, while the ship stood upright mirrored in her own stagnant image, and cans we threw overboard glinted in the blue depths, down and down and down, a hundred fathoms down, it seemed where the light still caught them. And our garbage littered our unseen wake, until the sharks came, and the sailors were afraid to swim, and the pitch bubbled in the seams of our wooden deck."

So wrote a young sailor of the terrifying ordeal of being becalmed in the doldrums, once the most fearsome part of the ocean, the part that has no wind.

the doldrums

The storms that speed across the shipping lanes of the Atlantic Ocean can bring torrential rain that will wash a man from the deck of his ship, the waves that pile together where oceans meet around the tip of Cape Horn can rise to 65 feet (20 meters) or more and the winds of a hurricane can reach speeds of 125 miles (200 kilometers) an hour, but for centuries perhaps the most hated spot on the ocean was the doldrums, where in the days of sail a thousand seamen died a slow, cruel death under the blazing equatorial sun. The name comes from *dol*, the Old English for dull, although in truth this part of the globe

Above: The proudest man-of-war in the British fleet was no match for the unpredictable violence of a storm at sea.

Above: A waterspout—a tornado that forms over water.

Without wind our beautiful Earth would be transformed into a cauldron of extremes ●

is anything but that, for there are other strong forces at work near the Equator. For while the doldrums are a place without wind, ironically they mark the point where that great global circulation may be said to begin.

Without wind our beautiful Earth would be transformed into a cauldron of extremes. The heat of the tropics would increase inexorably, while to the north and south the planet would freeze beyond the point at which life could survive. Instead we have wind, the circulation of air around the planet, ultimately mixing hot and cold, wet and dry, to create the rich confection of cloud and clear sky, rain and storm that make up the weather that we all accept as part of our lives. For centuries, as the ocean explorers who followed Christopher Columbus ventured further and further from the civilizations of Europe they came to learn of the strong, steady winds that would push them towards their destinations in the New World and the adventures that lay ahead of them. As they headed south away from Europe they learned to pick up the easterly winds that became known as the trade winds, which drew them on towards the Caribbean and the Gulf of Mexico. They learned to find a point near the Equator where they could cross it and pick up the easterlies of the southern hemisphere that would carry them on to the Spanish and Portuguese colonies of South America. Or, if they were further south they would ride on the steady westerly winds of the higher latitudes that would carry them on around the Cape of Good Hope to the spice lands of India and the east. The return from the Americas would rely on the steady westerly winds of the North Atlantic that would bear them on a quick, albeit stormy, passage home. These winds were the constants of the globe—stronger or weaker, they would almost always be there—but the doldrums marked the strange boundary near the Equator where, if a ship's captain ventured too far south, or tried to cross too soon, the wind would drop and the sails would flap unfilled.

The doldrums are a band around the Equator where the airstreams borne by the two sets of trade winds, in the northern and southern hemispheres, flow together and meet. Technically known today as the Intertropical Convergence Zone (ITCZ), this band fluctuates slightly away from the Equator and back according to the season. But wherever it lies, the point where the winds meet becomes a place of no wind, for where the airflows come together there is nowhere else they can go, except up. They lie above the region where the sun's heat is strongest on the globe, where the air warms and rises off the ocean in a still, moist column, lifting moisture from the sea as water vapor, climbing to form huge, towering cumulonimbus clouds. And where sometimes the intense heat is enough to lift waterspouts into the air, which drift aimlessly across the sea before collapsing as swiftly as they appeared.

But as this warm mass of air rises it begins an extraordinary journey. Slowly, over weeks or months, it becomes part of the global circulation of the atmosphere—and this circulation may take it as far as the North Pole and back in a cycle that sees a parcel of air sometimes at the edge of space, over the highest thunderclouds, sometimes whipping up waves in a surface wind, sometimes moving on to whirl in a tropical cyclone, sometimes falling to Earth as an icy snowstorm.

the cycle of the winds

If today the moods and swings of the weather seem mysterious, inexplicable and unpredictable, think back three or four hundred years and imagine what knowledge of the forces of nature was available to even the most intelligent and educated people of the day. The answer is very little. In the seventeenth century the thermometer was still the plaything of eccentric dabblers in science, such as Galileo's patron the Grand Duke Ferdinand II of Tuscany, who was possibly one of the least effectual rulers of his state but who was, fortunately for the rest of us, passionately interested in science. He developed one of the great astronomer's original ideas and turned it into a device that might seem more at home at a dinner party than in a meteorological station. Galileo had come up with the notion of using air's property to expand and contract with heat and cold to draw water up a closed glass tube as the air inside it cooled and contracted. It was a hopelessly inaccurate affair, because, of course, the air was also influenced by changes in atmospheric pressure, about which he knew nothing; but he used the device to measure what he called degrees of heat. Ferdinand replaced the air with wine, the expansion and contraction of which is not so affected by air pressure, and obtained a marginally better result. But it was not until the mid-1700s that the first reliable temperature scales were created by the German scientist Fahrenheit, who constructed a mercury-based thermometer for the first time, with the boiling point and freezing point of water coming out at 212° and 32° respectively. Anders Celsius, a Swedish astronomer, followed with a far neater proposal for 100° to be the difference between these two key temperatures, but it is not often reported that his first idea was to have the boiling point at zero, and freezing at 100°. It was only when this notion was reversed after his death that the scientific world latched on to Celsius as the universal temperature scale it has now become.

So the knowledge available to one Edmond Halley in 1686 was limited,

Above: Famed for his insights into astronomy, Galileo Galilei (1564–1642) was also an early experimenter in meteorology.

Above: As a young man Edmond Halley (1656–1742) undertook an ocean voyage to the Indies, and recorded his observations as the first meteorological map.

to say the least. Famous for his later reputation as Astronomer Royal, and particularly for the comet that bears his name, the young Halley is also of interest for being the creator of the very first meteorological map. The result of a voyage to the Indies, it is a simple chart of the tropical latitudes of the Indian and Atlantic oceans and was designed to map the trade winds and monsoons which, in those days, had a major effect on overseas trade. The trade winds had been discovered by Christopher Columbus on his first voyages to the New World; and their constant easterly flow had become the key to successful navigation and, ultimately, conquest. However, no one could explain why they blew the way they did.

Together with the map, Halley also published a paper which attempted to explain the trade winds, and he had clearly begun to understand why they were such a consistent force for navigation. A remarkable thinker and scientist, he hit upon an explanation that was very close to a perfect description of the engine of the global weather system. He suggested that the flow of the winds was created by the sun heating air at the Equator. He knew that air would rise when warmed, and that this would leave an area of low pressure below, so he reasoned that cool air would be drawn into the space left behind—and, on a global scale, cold air had to come from the poles. So at sea level the cool air was drawn in from the freezing poles, while high in the atmosphere the warm air at the Equator would move outwards towards the poles, in a continuous cycle of high-level wind blowing out and low-level wind (the trade winds) blowing in towards the Equator. It was simple and elegant, it was almost right, and it explained the becalming stillness of the equatorial doldrums, where warm air rose gently aloft.

However, even after his remarkable ocean voyage, Edmond Halley remained unable to explain why the trade winds always flowed from the east. Fifty years later an English barrister, George Hadley, who dabbled in science and became intrigued with Halley's work, came up with the answer: the rotation of the Earth. Any object flying through the air appears to be deflected from its path because, while the object moves in a straight line through space, beneath it the Earth is rotating relentlessly about its axis. Hadley suggested that because the surface of the planet rotates with the greatest speed where its circumference is greatest, at the Equator, air arriving from the poles would have much less rotational speed and lag behind the Earth's fast-moving surface as it neared the Equator, and would thus appear to be arriving from an easterly direction. So the wind heading towards the Equator is deflected to come from the east: the easterlies. This phenomenon is known today as the Coriolis effect (see Chapter One). The huge overturning of air, suggested by Halley, became known as a Hadley cell, but 150 years later this simple model was refined as meteorologists discovered that there were, in fact, three bands of air, three cells, that encircled the globe to the north and the south of the Equator, and that the air moved from one to the other as it worked its way out to the poles and back.

Right: Edmond Halley's weather map of the trade winds and monsoons. Notice that his symbols for wind direction were shaped like comet tails.

GLOBAL CELLS

It was a US Navy officer called Ferrel who worked out that Hadley's and Halley's simple model of a continuous cycle of air from Equator to pole and back could not be correct, because in the mid-latitudes the surface winds were not easterly, as would be expected from air moving towards the Equator. Instead there were prevailing westerly winds that the sailors of the expanding European empires had used on their long homeward journeys from the Americas. And there was also another horrific area of calm, known as the "horse latitudes" because desperate *conquistadores* were said to have eaten their horses or thrown them overboard to conserve dwindling supplies of water. This region occurred at about 30 degrees north of the Equator (south in the southern hemisphere), and suggested another break in the global circulation pattern of the wind.

What Ferrel proposed was a three-cell system. Taking the northern hemisphere on its own (for simplicity), what happens is that the mass of warm air high above the Equator moves north, converging on the pole, but it does not reach there. As it moves north, the narrowing of the globe means the air is squeezed together and builds up its density so that it sinks down to the surface, at the horse latitude of 30 degrees or so. From there, some of the air moves south again to create the trade winds, deflected by the Earth's rotation to become easterlies. But from the horse latitude some of the air moves north at the surface, this time deflected to the right by the Coriolis effect, and thus creates the reliable westerlies. So now there is a second cell, with the wind at the surface flowing generally towards the pole from latitude 30 to about latitude 60 degrees. There it meets cold air flowing down from the pole, at a boundary called the polar front. The converging air rises and creates storms, but high above some air moves back towards the boundary at the horse latitude again, completing the second cycle. Finally, north of the polar front, the high-altitude air flows towards the pole, where it finally descends to begin the long journey south.

These three cycles are known as the Hadley cell, the Ferrel cell and the polar cell, and they are mirrored exactly in the southern hemisphere. The air heated at the Equator does indeed move to the pole, though not in one simple continuous flow, as Edmond Halley believed, but in the three cells, making its way alternately high in the atmosphere, low down on the Earth's surface and high again, before returning to the Equator. Over a century of scientific thinking had

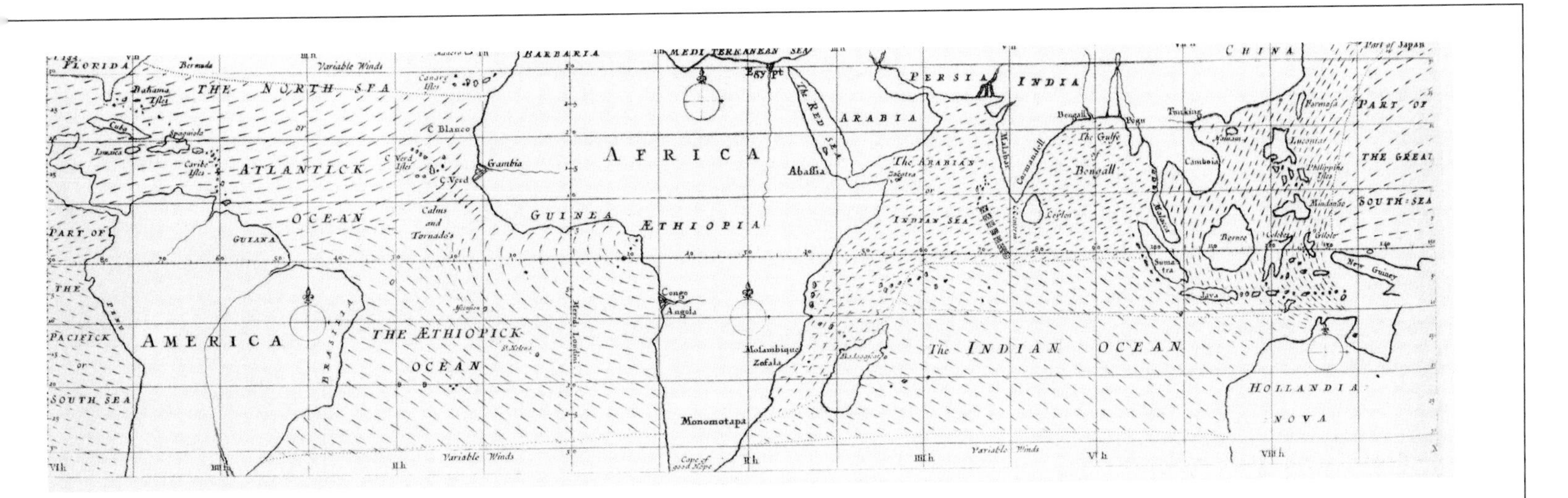

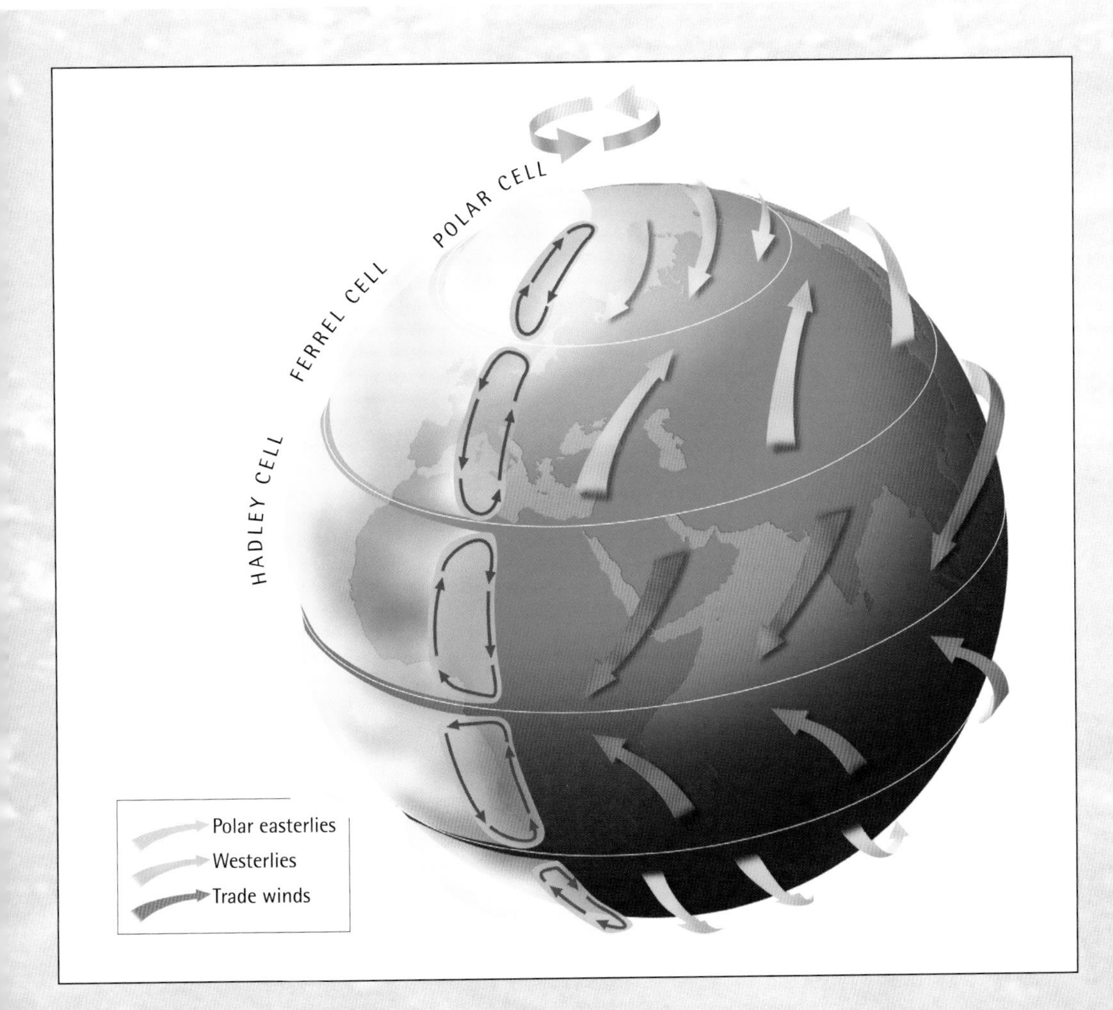

finally arrived at a simple explanation of the thrills and terrors of navigation at sea.

As with all large and apparently complex phenomena, the huge forces of nature that drive our weather had been reduced to a beautiful but powerful system. The sun's heat, and the constant need for hot and cold to even out across the globe in this simple cycle of hot-to-cold-and-back-again, are what create the constant movement of winds over the planet.

Left: The global movement of air from the Equator to the poles is divided into three cells in each hemisphere at latitude 30 and 60 degrees. The surface winds beneath each cell adopt a prevailing direction: the easterly trades; the westerlies; and the polar easterlies.

birth of a typhoon

A tiny parcel of air, breathed out of a sailor's lungs, or swept into motion by the flap of a sail, can so easily become part of the remarkable disturbance that creates the most powerful force in the atmosphere. In that quiet region near the center-line of the planet, the Intertropical Convergence Zone, the bands of moist air, driven by the trade winds moving close to the surface of the ocean, are steadily converging. Where two such bands of air meet from different directions they begin to mix. Passing over water heated by the equatorial sun, the air becomes warm and very moist, rich with water vapor from the ocean. Caught between two giant streams pressing together, its only outlet is to rise.

As it rises the air encounters lower pressures, and so it expands and cools. This makes the moisture in the air condense out to produce clouds and this process releases heat that makes the air at the surface expand, thus forcing more air to rise. As more and more clouds form thunderstorms begin to develop, and as the air rises, more is drawn in at the base to replace it—so the cycle intensifies. This is the basis of a tropical disturbance, and between June and November every year the Atlantic Ocean sees about ninety of them, one every couple of days. Only a quarter of these disturbances develop further, but that's where the fun really starts: if more and more air is able to rise, it marks the beginning of what is known as a tropical depression. And the critical factor that determines whether disturbances develop further into depression, or merely fizzle out, must now come into play. Our parcel of air will either move on to undergo the roller-coaster ride of a lifetime, or it will dissipate back to the sea. The key is the wind flowing above it. As the air climbs, it meets winds that are already flowing in layers at different altitudes. If these layers of wind are blowing from different directions or at different speeds the rising air-column will be broken up and the fledgling thunderstorm will be torn apart before it has a chance to form. But if the winds are uniform, the moist air rises unchecked, condenses and releases more heat. In turn this warms the air at high altitude, which expands and creates high pressure as the warm molecules of air move faster and exert more force on the other molecules that surround them. The expanding air then begins to move away from the developing thunderstorms and thus draws even more warm air up from below it.

Nature has certain minimum requirements for a tropical depression to develop into something more sinister: the sea needs to have a surface temperature of at least 80°F (27°C) and be warm to a depth of 230 feet (70 meters). That is the sort of stored-up energy needed to drive the monster that is about to form, a creature that requires vast quantities of warm moist air to stay alive and which generates enough power to provide the total electricity requirements of the USA for half a year. Such conditions are not found everywhere across the globe, but they are readily available amid the warm tropical waters of the Atlantic and Pacific oceans just close to the Equator, where there is a seemingly limitless supply of water vapor to rise and feed the thunderstorms forming above.

By now our little parcel of air is caught up somewhere among the billowing clouds that are so indicative of thunder. But for the innocent group of thunderstorms to develop into something more threatening, another ingredient is needed: the Earth's spin. The Coriolis effect means that the rotation of the Earth deflects the wind and twists it to the right in the northern hemisphere, and to the left in the southern hemisphere, and the further north or south you go, away from the Equator, the stronger that effect becomes. In the summer months the changing pattern of solar heating of the Earth pushes the equatorial trough on which these thunderstorms develop slightly away from the Equator. How far it shifts is critical. At least 5 degrees latitude above or below the Equator and the Coriolis effect is strong enough to create a tropical depression, and the little group of thunderstorms that is forming begins to rotate. This spinning effect forces the thunderclouds into a series of organized bands, wrapping them around the point of low pressure at the center. The winds get faster and faster towards the center, producing a spiral structure. It is now beginning to take on the look and feel of a hurricane.

Above: A tropical typhoon building up in the western Pacific, seen from the space shuttle. The eye of the storm is clearly visible.

us naval disaster story

A typhoon was the force that bore down on Admiral William F. Halsey and nearly a hundred ships of the US Navy's Task Force 38 in December 1944. Their mission was a bold one: to support a full-scale invasion of Luzon, the most northern island of the Philippines, which was still in Japanese hands. On December 17, the ships were refueling in the Philippine Sea in preparation for the assault, but the swell was rising and the winds were blowing a harsh 35 miles (56 kilometers) an hour from the north-east. It was the start of two days of hell. Halsey called on his meteorologist on the flagship to provide a forecast that would enable him to seek out calmer waters to complete the dangerous refueling, which by now was unmanageable with fuel lines being torn from the nozzles of the tanks as the huge ships bucked and rolled.

The meteorologist had very little firm data on which to base his judgement, as there were no networks of weather stations, no regular reports from pilots in the area, and the only source of weather maps was far off in Pearl Harbor on Hawaii, which itself had little data on the hostile regions at the fringe of the war zone in the Pacific Ocean. But storms develop and move in that part of the world according to certain principles and this enabled the meteorologists to give a confident assessment of the weather and a firm prediction of how to avoid it. The fear, of course, was of a typhoon. The path that these storms normally take across the Philippines is to the north-west. But the waves that were tossing the ships were coming from the north-east and it was well known that the swell running ahead of a typhoon normally moves in the same direction as the advancing storm. And there was another factor: because the winds of a typhoon spiral around its center, they are normally first felt in a direction at right angles to the direction of the approaching typhoon. So the direction of the ocean swell and the direction of the winds both pointed to something other than a typhoon, and the admiral was confident of his meteorologist's assessment that the cause was a simple storm far to the east of the task force.

High overhead was a cold front running from north-east to south-west and everything suggested that when the storm eventually did arrive it would hit the front and move off to the north-east, so the fleet turned to the north-west to put the protective cold front between them and the storm they assumed was to blame.
In fact turning to the north-west meant they were running ahead of a violent typhoon, which was now far behind their stern. But as the ships were moving faster than the storm, for a while the weather improved. The following day, persuaded by the calmer weather that the storm had been diverted by the cold front as predicted, the fleet turned tail and sailed back towards the south-east to regain their position for the invasion—but also straight back towards the impending power of the typhoon. By mid-morning it was upon them.

As they slid into the deadly storm, its speed increased to 125 miles (200 kilometers) an hour, whipping up waves over 80 feet (25 meters) high. For the men on the ships there was no escape: they could only hope that their vessels would ride out the storm. Many of the task force vessels escaped the full anger of the typhoon, for by now the fleet was widely dispersed, but some ships sailed directly into the center of the storm, and of those some never returned. The large aircraft carriers and battleships were safe, but at the other end of the scale the small escort carriers and the sleek, high-speed destroyers were in serious difficulty. On board the escorts, airplanes tore loose, slid across the decks and burst into flames, and the crew risked their lives

Above: Admiral William F. Halsey faced a court of inquiry after sailing his fleet into the heart of a tropical typhoon.

to tip aircraft overboard to save the ships. The slender destroyers, designed to slice through the sea at speed, could not maneuver in the high seas and three of them sank. Men below decks had to struggle to open hatches against the force of the wind, only to find themselves immediately washed overboard; even those with inflated life jackets were crushed by the waves against the sides of the ships. The captain of a stricken destroyer recalled later how "the force laid the ship steadily over on her starboard side and held her down in the water until the seas came flowing into the pilothouse itself. The ship remained on her starboard side at an angle of 80 degrees or more, as the water flooded into her upper structures." With the ship keeled over in this position, in the end the captain had no choice but literally to step from his bridge, straight into the sea.

Rescue planes began to arrive three days later, when the typhoon had moved on and blown itself out. In all 150 aircraft were lost, and from the three destroyers sunk, 790 men were drowned, leaving only eighty-two survivors. The task force had been so badly damaged that it could not join in the attack on the Philippines. Sadly for Admiral Halsey, who was blamed for the handling of the disaster, he faced an almost identical chain of events six months later when, charting a course to escape another typhoon, he sailed straight into it. Of his force, which was then preparing to attack Okinawa in Japan, thirty-three ships were damaged, seventy-six planes lost and six men killed. With the Philippines typhoon, the US Navy had suffered one of its worst known disasters and the urgent need to predict these killer storms became ever more pressing.

The universal meteorological term for these violent storms that start life over tropical waters is "tropical cyclone," but they have different names around the world: a 'typhoon' in the northern Pacific and a "cyclone" in India and Australia are exactly the same phenomenon, and in the Atlantic it has the more familiar name of "hurricane," which originates from the Central American Taino word *huracan*, meaning "god of evil."

Ironically, early radar equipment aboard Halsey's ship had revealed a clear profile of the typhoon that threatened him, but so little was known about either the storm or the radar technology that no one had been aware of what they were seeing. By the end of the war, however, reconnaissance missions were being flown regularly into the typhoons of the Pacific and the hurricanes of the Atlantic, and hurricane-hunting had been born. The first flight into the center of a hurricane had occurred only in 1943, carried out by a US Army Air Force pilot who had been determined to test the efficacy and universal application of instrument flying. In penetrating the turbulent spiraling winds, and crossing into the calm eye at the center, he revealed the extraordinary range of conditions that exist throughout a hurricane. And just to prove that it was a safe thing to do, after landing and describing the most extraordinary flight of his life, Colonel Joseph Duckworth promptly offered the station's weather officer a chance to take a look—and flew through the hurricane a second time.

TYPHOON/HURRICANE SCIENCE

A hurricane is defined as an intense storm of tropical origin, with sustained winds exceeding 64 knots—about 74 miles (120 kilometers) per hour—and today we know much more about what goes on inside its heart. Satellite images have now made us familiar with an almost benign and certainly beautiful view from space that reminds us of a spinning Catherine wheel, or some kind of fluffy, rotating, screen-saver device working its way across the globe. But the beauty of the spiral of cloud, with its black dot of an eye at the center, belies enormous power and commands respect. The average hurricane is about 340 miles (550 kilometers) across—approximately the length of Scotland. The eye in the middle is an area of 12–30 miles (20–50 kilometers) diameter which has thin, broken cloud and light winds, and patches of clear blue sky. The spiraling clouds that blow towards the center of the storm are called "spiral rain bands;" they twist round the eye, always in the same direction—anticlockwise in the northern hemisphere, clockwise in the south. The warm, moist, tropical air flows ever inwards, where it builds up in a circle and rises, condensing out to create a ring of intense thunderstorms tightly packed around the eye in the middle and extending up to 9 miles (15 kilometers) in height. Known as the "eye wall," this is home to the heaviest rain and the strongest winds, which can rise to well over 125 miles (200 kilometers) an hour. At the top of the thunderstorms that make up the eye wall the air has lost most of its moisture and flows outwards high above the maelstrom below, cooling as it goes. Ironically this outflowing air creates winds that blow in an anticyclonic direction (clockwise in the northern hemisphere) at a great distance from the storm, so several hundred kilometers away the cooled air gently sinks and heats up, bringing with it clear skies—a deceptive precursor to the storm that may be approaching from far away, and just the kind of sign that Admiral Halsey's fleet misinterpreted to such disastrous effect.

Meanwhile, back at the center of the storm, the violent thunderstorms of the eye wall release large quantities of heat, warming the air and so building up even higher pressure above, forcing some of it down inside the eye, creating the calmer conditions and absence of winds at the center. If you could ride inside a typhoon, you would see whirling rain clouds, with ice crystals forming at the top, and feel the heat and hear the incredible sound of lightning crackling around you. One of the early reconnaissance pilots recalls the extremes that are experienced on the tumultuous flight through a hurricane:

"One minute this plane, seemingly under control, would suddenly wrench itself free, throw itself into a vertical bank and head straight for the steaming white sea below. An instant later it was on the other wing, this time climbing with its nose down at an ungodly speed. I stood on my hands as much as I did on my feet.

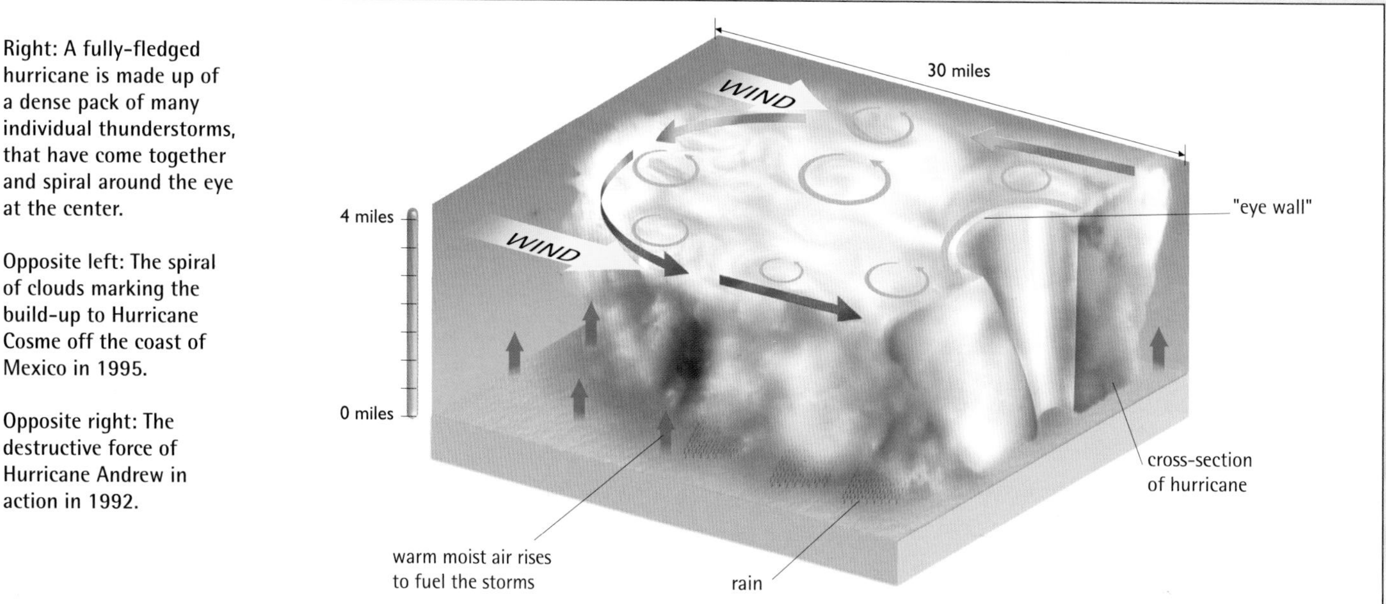

Right: A fully-fledged hurricane is made up of a dense pack of many individual thunderstorms, that have come together and spiral around the eye at the center.

Opposite left: The spiral of clouds marking the build-up to Hurricane Cosme off the coast of Mexico in 1995.

Opposite right: The destructive force of Hurricane Andrew in action in 1992.

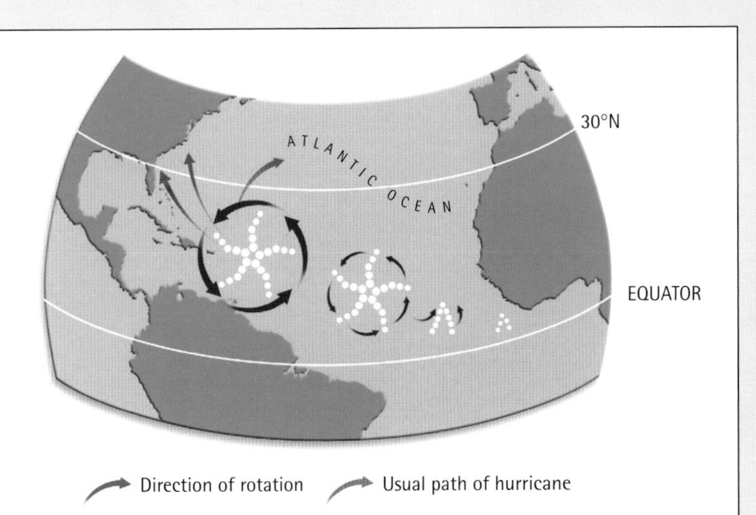

Right: Hurricanes are born off the coast of Africa, when a line of thuderstorms begin to spiral as one mass, due to the Coriolis effect of the spinning Earth. They track across the ocean to hit the east coast of the Americas with full force.

Rain was so heavy that it was as if we were flying through the sea like a submarine. Navigation was practically impossible. For not a minute could we say we were moving in any single direction—at one time I recorded 28 degrees shift, two minutes later it was from the opposite direction, almost as strong."

But once a pilot had penetrated the extreme violence of the eye wall, the eye itself afforded a place of calm in which it was possible to circle a plane, radio back weather reports and even take a coffee break. Bizarrely, a pilot might meet flocks of birds that had become caught in the eye of a hurricane as it formed, now unable to fly out to where they had intended to travel. Instead they were forced to journey the ocean inside a huge natural cage until the storm finally faded and released them.

A fully fledged hurricane is a vast, self-sustaining engine, a hundred times larger than a thunderstorm and a thousand times more powerful than a tornado. An ordinary summer thunderstorm can have the power of three nuclear bombs; a hurricane has twenty-five thousand times that power, and if it remains over warm ocean water it can be fueled for days. But the Coriolis effect is also responsible for hurricanes eventually fading out. The effect of the spin of the Earth means that hurricanes and typhoons generally track westwards, and after they have spent days sweeping the breadth of the Pacific or the North Atlantic Ocean, their direction of travel inevitably takes them over cooler waters, or finally over dry land. There they simply run out of fuel—a lack of heat and moisture means that fewer thunderstorms can be formed and thus the temperature at the top of the system is lowered. This effectively puts a lid on the storm—the air stops expanding and moving outwards from the top. Air from the ground will not rush in, because none is being lost higher up, and so the winds subside and the hurricane downgrades to a tropical storm. It may travel on for weeks as a mid-latitude depression before finally all trace of it is lost.

Left: The south Florida coast bracing itself for the onslaught of Hurricane George.

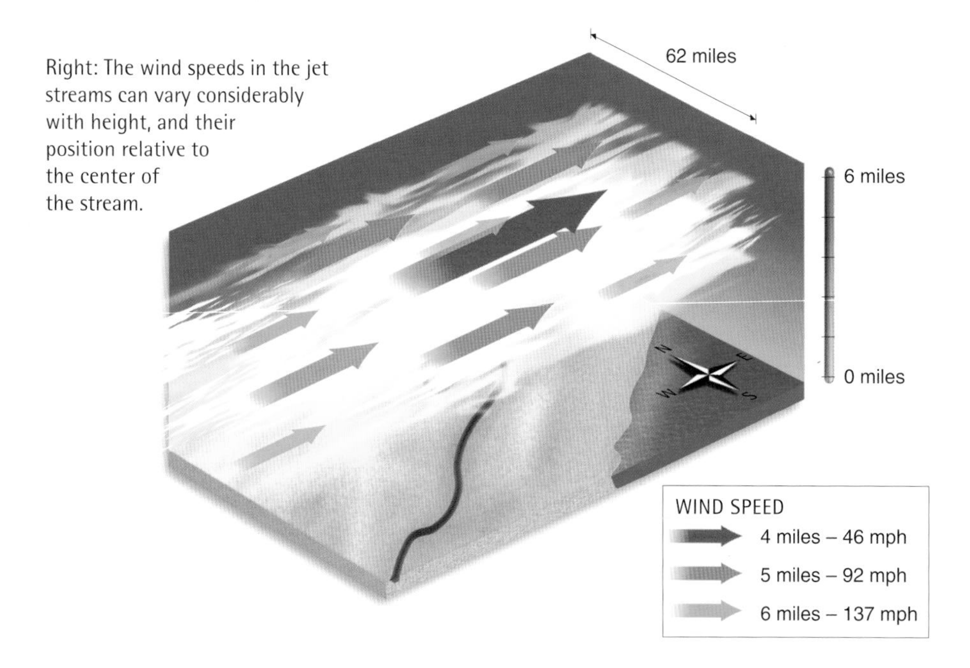

Right: The wind speeds in the jet streams can vary considerably with height, and their position relative to the center of the stream.

Above: The line of the subtropical jet stream passing over Egypt is

jet streams

Hurricanes blow themselves out as they reach the edge of the first of the three cells that move warm air towards the poles: at the boundary between the Hadley cell and the Ferrel cell. But that boundary creates a new and even more powerful flow of air, not near the ground but high in the atmosphere at the upper limits of what we call the weather. Huge volumes of air of different temperatures are rising and falling at the point where the two cells meet and the boundary shifts constantly, waving snake-like, north and south, as one cell expands against the other, only to contract as pressure builds on the other side. This is not a fixed line around the world but, if it could be made visible, would be more akin to an ever-changing battleground where one army pushes polewards with reinforcements of warm air, only to be repulsed by greater quantities of cold pressing the other way. The result is to create huge differences in pressure in the air aloft, and just as any wind on the ground is created by differences in pressure between blocks of air, so very strong winds build up at the top of the boundary, creating fast-moving rivers of air that encircle the planet. These are the jet streams, currents of air typically hundreds of kilometers long, 60 miles (100 kilometers) wide and over ½ mile (a kilometer or more) deep, which can reach speeds of 200–300 miles (300–500 kilometers) per hour. There is a jet stream at each of the four boundaries between the global circulation cells, and they have a profound influence on the weather that develops beneath them.

japanese bombs

It was really only during the Second World War that the power of the jet stream became apparent, but since that moment it has proved to be a force that has repeatedly taken humans by surprise. One of the first shocks came in December 1944, in the north-west USA. Two logging workers in the forests of Montana stumbled across the remains of what was clearly a large balloon, tangled in the trees. It appeared to have simply arrived, silently, in the middle of the night. The balloon was large enough to have carried people but there was no gondola in sight.

-ked by streaks of very high-altitude clouds.

What made it worrying, though, was that printed on the thin fabric spread out in the branches was the sign of the rising sun, the national emblem of the Japanese. FBI and army investigators hurried to the site, and the whole thing was quickly hushed up until they could establish what had happened. When the story eventually broke the media conjured up visions of Japanese spies parachuting in from above and now roaming the countryside, but it turned out that the balloon, which even had its date and place of manufacture in Japan printed on its canopy, had carried a simple bomb. Over that autumn other reports had emerged of fireballs mysteriously appearing, of balloons that exploded and of forest fires being ignited. These events occurred with no loss of life, up and down the western states of the USA from Colorado to California, until a party of children in Oregon on a Sunday-school outing to the seaside began to investigate a strange object they found in the sand. It exploded, killing five of them and their teacher.

The clever construction of the balloon bombs was eventually established when one was captured intact and herded to the ground by an American Air Force pilot. But how had they reached the USA? Balloon experts in America realized that they could not have traveled the distance across the northern Pacific from Japan without some considerable help. The most likely solution was that a fleet of submarines was launching the bombs from a short distance off the American coast. However, after the war the truth was revealed to have been beautifully and ingeniously simple: they had flown in the jet stream. The Japanese had discovered the high-level band of fast-flowing air that could enable their planes to travel very quickly towards the coast of the United States, shaving an astonishing six or seven hours off the flying time. It was a one-way flow, but perhaps the notion of suicide missions was such that little thought was given to the long struggle such aircraft would face if they wanted to return. Either way, the really clever idea that the Japanese high command came up with was the "Fugo bomb," and it turned out that some nine thousand of them were launched as an offensive against the USA from late 1944 onwards. How they worked was ingenious: large and lightweight, they were constructed of thin ricepaper and filled with hydrogen, which itself would burn violently when the bomb went off. An incendiary bomb hung below the canopy, and on board was a device that worked out the altitude of the balloon from the external pressure of the atmosphere, which gets less and less the higher you climb. At the height where the balloon should reach the fast-moving jet-stream winds, a small valve was triggered to open and release some of the hydrogen, in order to prevent it from rising further. Thus the wind would carry it on its way. If the balloon fell too far, its little altimeter would instead trigger the release of a small sand bag. Unmanned, silent and sinister, the balloon would rise and fall, keeping neatly within the jet-stream, and be blown at high speed to the USA to create terror.

Barely a thousand Fugo bombs ever reached the USA during the war, and apart from the children in Oregon there were no casualties. Nevertheless, the surprise and anxiety caused by the fact that fire could simply fall out of the sky onto people who were far from the theater of war made the bombs a surprisingly powerful reminder of the potency of the Japanese threat.

Ironically, the same winds that provided such an opportunity for Japanese attack on the USA also posed a huge difficulty for American pilots trying to hit back. The giant B-29 bombers, the "Superfortresses" that brought waves of destruction to Japan at the close of the war, were able to fly so high that they frequently found themselves caught out by the high-speed winds. On one mission a squadron of ninety-four bombers flying at around 30,000 feet

(9,000 meters) passed over their target and turned east to make the bomb run, but found themselves blown so fast that the bomb-aimers could not get a fix. On another mission a headwind of 175 miles (280 kilometers) per hour slowed a B-29 to such a degree that it became a sitting target for the anti-aircraft guns below. Whatever the military failure of the mission, it provided a new insight into these rivers of air that flowed at high speed around the world and which no one had ever imagined were there. After the war the US government began to realize the potential value of understanding the jet streams and serious research was launched into their characteristics, position, speed and predictability. Investigations by high-altitude balloon flights confirmed their existence. Some exceptional ones were up to 4 miles (6 kilometers) deep, anything up to 500 kilometers (300 miles) wide, and moving at speeds of up to 300 miles (500 kilometers) per hour. And, taking a leaf out of the Japanese book, the US military used them to carry high-altitude balloons bearing spy cameras around the world, over the USSR, and back home.

Left: A computer-generated image of Japanese "Fugo bombs" carried by balloons. Borne along in the jet stream, they were an attempt to terrorize the USA in the Second World War.

Above: Few of the bombs ever reached their destination.

MISSING

Ever since the Second World War the jet streams have continued to put life at risk, and even take it. As recently as the turn of this millennium it has been revealed that a jet stream was almost certainly at the heart of one of the longest-standing mysteries in aviation. In 1947 a British Lancastrian airliner set out from Buenos Aires in Argentina to fly to Santiago in Chile—not a great distance, a routine flight but a high one as the mighty Andes mountains stood between the aircraft and its destination. The plane was a converted Lancaster bomber named "Stardust," and on board were the pilot and his navigator, both highly experienced, three other crew and six passengers who included a German emigré carrying the ashes of her dead husband to Chile, a Palestinian businessman with a diamond sewn into the lining of his jacket and a king's messenger bearing diplomatic papers. For all of them the flight was to turn to tragedy, because the plane simply disappeared.

Forty-five minutes before landing in Chile, the Stardust should have crossed the Andes close to Aconcagua, the tallest mountain in South America, and shortly before its time of arrival it duly reported in morse code to Santiago that it was just four minutes from touchdown. But the plane never arrived. Search parties flew out from Santiago, tracing the route of the aircraft, ranging wider and wider across the mountains, but no sign of it could be seen. Then in January 2000 a climber on Mount Tupangato in the Andes, 50 miles (80 kilometers) from Santiago, came across an old, mangled Rolls-Royce engine lying out on a glacier—and it belonged to Stardust. Spread out across the ice nearby were the frozen remains of some of the people on board. This was an area that had been thoroughly searched at the time of the accident and, although the plane had reappeared, the mystery of how it came to be so far from its reported position, on the wrong side of the Andes, and why it had remained hidden for so long, only deepened. The Argentinian army mounted a full-scale expedition to examine the site, and air-crash investigators carefully reconstructed the events of the fateful flight. Painstaking analysis of the wreckage out on the freezing mountain revealed that the Stardust had been in full flight when it met its end, but why had it landed on this glacier when it should have been well clear of the entire mountain range? Consulting a glaciologist produced the first clue: it became clear that the aircraft had not crashed where it was found, but much further up the mountain. It seemed that Stardust had struck the sheer snow-covered face of the mountain, possibly dislodging an avalanche in the impact, so that the wreckage had been completely covered in snow and ice. The remains of people and plane had become entombed in the layers of ice that made up the glacier and were carried downhill and finally released to the surface as the ice melted on the warmer lower slopes—to be rediscovered over fifty years later.

While the disappearance was now explained, the cause of the crash

Opposite: The discovery of a Rolls-Royce engine on a glacier in the Andes revealed clues to the fate of the Stardust.

Above: The airliner Stardust was one of a fleet of Lancaster bombers converted to carry passengers after the Second World War.

remained a puzzle. The crew were experienced, and an 50-mile (80-kilometer) navigational error on this short flight needed an explanation. But a study of the weather reports of the time, and of the radio log of the air-traffic controllers, has revealed the likely reason for the disaster. Conditions were bad in the mountains and the crew had radioed for permission to climb to 24,000 feet (7,300 meters) to get above the clouds. As they were now also high enough to fly directly over the mountains, almost certainly they turned to head directly for Santiago, reckoning to descend when they had got to the other side of the Andes. But at that height they would have encountered the winds that they knew nothing about: the jet stream. The prevailing weather conditions on that day were, we now know, ideal for the development of the jet stream at the latitude of the flight. Traveling at over 93 miles (150 kilometers) an hour in the opposite direction, the high-altitude torrent of air would have reduced Stardust's ground speed to a snail's pace. But the crew had no idea it was happening. Believing themselves to be well over the mountains, they would have relaxed as they descended into the clouds towards what they thought would be a conventional landing in Santiago. Instead they flew straight into a solid wall of rock.

ill wind

The air that began its journey over the doldrums, soared high aloft inside a hurricane and rushed along in the high-speed rivers of the jet stream above, now travels across the Pacific on the historical heels of the Japanese Fugo bombs and finds itself at the boundary of the next wind-circulation cell to the north. This is the Ferrel cell, and it covers the mid-latitudes where the bulk of the world's western population lives.

Far below the jet stream winds, over the western land mass of the USA, another draught of air is beginning to develop: the hot, dry, malevolent chinook wind. The prevailing surface winds of the mid-latitudes, in the Ferrel cell, are the westerlies. As they sweep in across the western USA they run into that dramatic wall of land known as the Rocky Mountains, which, together with the Cascades to their north, form an effective barrier between the Pacific coast and the Great Plains and prairies on the other side. And the eastern side of the Rockies is the home of the chinook wind. As dry air piles up high over the top of the mountains, a trough of low pressure is created on the eastern side, which draws the air down the slope. As it falls, it accelerates and gets rapidly squeezed against the air and earth below: it is this compression that warms the wind. Indeed the chinook is also known as the "snow-eater" for it can melt 12 inches (30 centimeters) of snow in a day. Folklore tells of a sleigh-ride where the traveler tried to outrun the chinook and found the front of his sled in snow, while the rear of the runners grated on soil for the whole journey—unlikely, but an indication of the remarkable ability of the chinook to conjure up stories of extremes. One of its quite remarkable documented abilities is to generate rapid temperature changes. When it rushes down to meet a basin of cold air below, it pushes down into it, only to find that the cold air rebounds against it in turn. This slopping back and forth once created a wild fluctuation of temperature in South Dakota, where an early-morning temperature of -4°F (–20°C) rose to 54°F (12°C), dropped back down to 12°F (–11°C) and then rose back up to 55°F (13°C) within the course of a couple of hours.

There are fast, falling winds with similar characteristics the world over, and different peoples have different names for them. In the Alps of Germany and Switzerland the wind is known as the Föhn, and its speed can cause terrible damage. In winter it may be useful in clearing snow, but at other times it can pick up dust and gravel and throw it at cars and windows to devastating effect. Perhaps the most notorious of the world's falling winds is the Santa Ana, where air from the high Mojave Desert in California is squeezed through the San Gabriel Mountains and shoots down across Los Angeles on the other side. It is heated again by compression, whereupon it plays havoc with the population. The sea off Long Beach is whipped up into foam, and gales of up to 60 miles (100 kilometers) an hour buffet the whole county. On one occasion 252 oil derricks were destroyed in one blast, and on another 15 million tonnes of dust fell on Burbank, home of the movie studios.

biometeorology

The Santa Ana is a violent and abrasive wind in more than a physical sense. To the population of Los Angeles it brings an increase in ulcer perforation, embolism, thrombosis, hemorrhage, migraine and electrical failures, and lowers industrial productivity and milk yield in cows. But it also brings far worse.

"There was a hot desert wind blowing that night. It was one of those hot dry Santa Anas that come down through the mountain passes

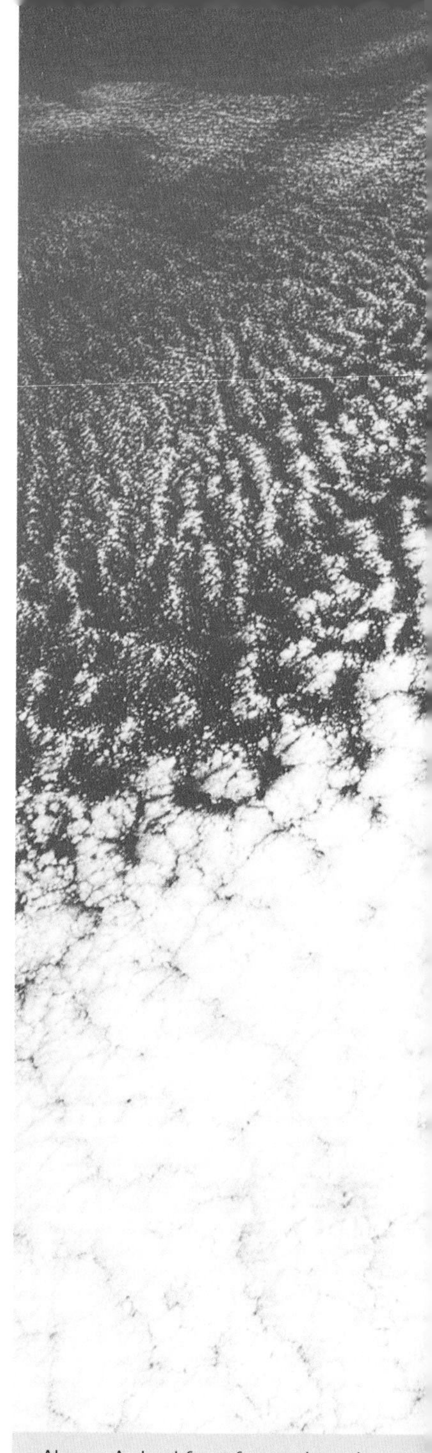

Above: A cloud front forms where the

warm air of the Santa Ana wind meets cooler ocean air off the Californian coast.

and curl your hair and make your nerves jump and your skin itch. On nights like that every booze party ends in a fight. Meek little wives feel the edge of the carving knife and study their husbands' necks. Anything can happen."

This classic piece of Raymond Chandler writing in a Philip Marlowe short story expresses perfectly the more extreme effect that winds like the Föhn and the Santa Ana are believed to have on the people who live in their path. A study in 1968 by Willis H. Miller gave a hint that the wind might indeed have a direct effect on the rate of killings in America's murder capital. He extracted the homicide data for 1964 and 1965 and compared the murder rate on days where the humidity was extremely low (a sign of the Santa Ana) with the normal murder rate for the city. He found a significant result. On average there were 14 per cent more murders on Santa Ana days, and during one wind, which blew for six days late in 1965, 47 percent more homicides were committed. Of course a single study from thirty years ago does not prove a causal link, but it is an indication that something might be going on. Certainly in the early days of California, when the justice system was perhaps more flexible, the defence of an extenuating Santa Ana wind was often heard in cases of passionate crime.

So the Santa Ana drives people crazy—but why? To try to answer that question, and others like it, the little-known science of biometeorology—the study of how the weather affects our health—has developed over the last thirty years. Thus far it has proved very difficult to pin down the causes of such effects on the physical and mental well-being of people in the grip of unusual weather, but some remarkable clues are emerging. Migraines are commonly claimed to be brought on by changes in the weather, and a recent piece of research on the effect of the chinook wind in Canada, which has the same mechanism as the Santa Ana and the Föhn, has revealed that a significant number of migraine sufferers had their headaches triggered on chinook days. But the surprising fact was that a different group among the patients found their migraines worsened on pre-chinook days, rather than on the day of the wind itself. So the link with the wind is there, but the biological mechanism that causes it may turn out to be quite different in different people. One of the atmospheric phenomena always associated with Föhn winds is an electrical activity called 'sferics, short for "atmospherics"—low-frequency electromagnetic waves, between 4,000 and 50,000 cycles per second, of "atmospheric origin." The background level of such waves is perhaps around 10,000, but with the approach of a storm can increase to 650,000. They are associated with thunderstorms, but also with large bodies of cold, unstable air, which are the precursors to Föhn or Santa Ana winds. More recent research has demonstrated a clear link between very low-frequency 'sferics and electrocortical activity in headache sufferers. Their alpha and beta brain waves were stimulated by the 'sferics and continued to be active for twenty minutes after the waves had been turned off. And extra low-frequency oscillations of atmospheric pressure have been found to have an effect on people's mental ability: their attention span, their memory and their mental processing.

While the evidence for a real biological and physiological connection to the weather is compelling, there is no clear mechanism, but a favorite theory for over thirty years has been the notion of positive ions in the air. And it's worth taking a more lateral view of the air around us to put that in perspective, as Lyall Watson does to brilliant effect in his provocative book *Heaven's Breath*. Air looks clean, and on a fine day far off in the countryside it feels fresh and good to breathe. But even the cleanest air on the planet probably contains some 200,000 stray particles in the half-quart of an average inflating lung. Wind is as much of an ecosystem as is the land or the sea. Under the microscope the air is actually a teeming mass of little particles; several million per lungful. Standing on a busy city motorway, that count can rise to perhaps 375 million particles. Of these, some 99 per cent will be minute particles of salt, clay or ash from forest or industrial fires and unsaturated hydrocarbons

Above: The Föhn wind is a fast-moving, warm wind that whistles down off mountains with surprising

produced naturally from vegetation. As a collection of trillions of particles in the air, they scatter light and form the blue and gray hazes of dry days. In addition there are larger fragments, blown up from the ground, carried from distant volcanic eruptions or even the remains of burnt-out meteorites; some of the particles are of paint, asphalt and rubber produced by the cars on the roads below. And then, living within this mêlée of material, can be found viruses, bacteria, spores of fungi, molds, mosses, algae and more.

With each breath of air we breathe we engage in a close and intimate relationship with the world around us, so it is perhaps not surprising in the least that our health is influenced by subtle changes in its composition.

suddenness and can whip up violent effects such as these.

Indeed, we are all aware of the risks of airborne infections and pollution. But in and among the soup of our lungful of air are a tiny quantity of things called positive ions, and it is these that doctors believe may be the cause of the ill temper brought on by certain winds.

Air is made up of molecules which are normally electrically in balance. Positively charged protons are countered by negatively charged electrons, and the molecule is stable—but not that stable. When, for example in a growing thunderstorm, air masses move against each other, or dust and water vapor mix with molecules of air, electrons can easily be knocked off the air molecules and attached to others. The result is ions: molecules that have either too much positive charge or too little—positive or negative ions. In the 1960s and 1970s a host of scientific research was done to find out if living things, including us, were sensitive to the presence of greater or fewer numbers of positive ions in the air. Seedlings of barley, oats and lettuce, even cucumbers, grew bigger in ionized air; bacteria were reduced in number, as were particles of dust and pollen—probably because negative ions tend to be 'sticky' and coagulate with other particles until they are too heavy to remain airborne. Mice even caught influenza more easily when ions were depleted. One of the major centers of scientific research into this field has been in Israel, and there, it seems, they have found that humans are sensitive to differences between positive and negative ions, and that it is the relative amounts of the two that are crucial. Too many positive ions building up in the air can result in breathing difficulties and lead to the release of serotonin, which is the most prolific and influential neurotransmitter in the brain. Serotonin is involved in almost all our moods and patterns of behavior.

The result of a sometimes uncertain but nonetheless growing body of evidence that ions can influence health has been the proliferation of sales of table-top machines to create negative ions that are "good for you." The marketing might have been easier if it was positive ions that were beneficial, but that's the way the evidence goes. As for the winds, the effect of the dusty desert winds that have prompted so much research in Israel is that they build up positive ions, because of the collision of air molecules with dry dust particles, while the stickier negative ions tend to coagulate with the dust and fall out of the sky. There perhaps is the explanation of the effect of an ill wind. Certainly in Europe, the effect of the Föhn on people is both accepted and acted upon. The German weather service offers bioweather forecasts on a regular basis, and in some Bavarian hospitals non-urgent surgery is postponed when the Föhn begins to blow.

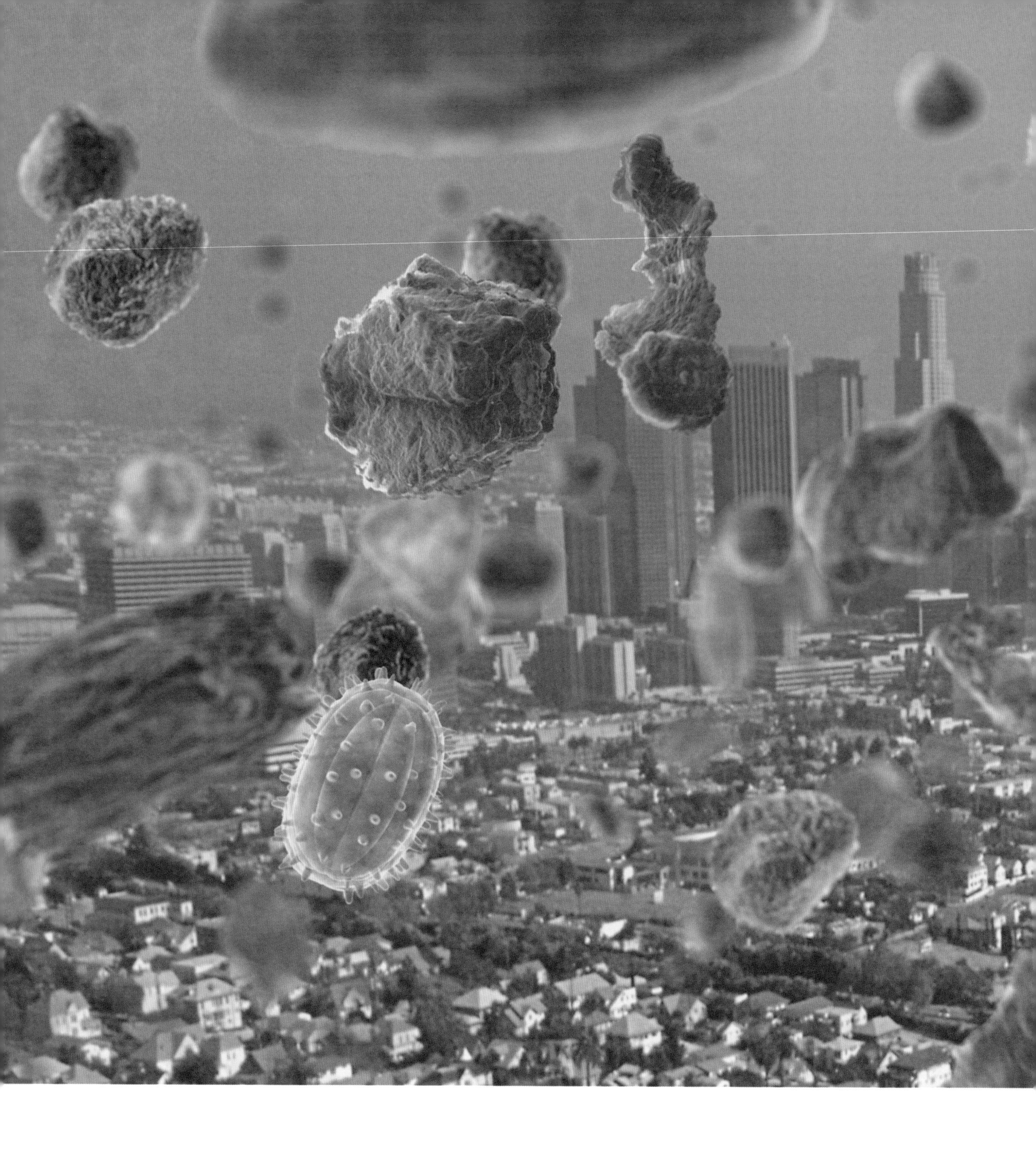

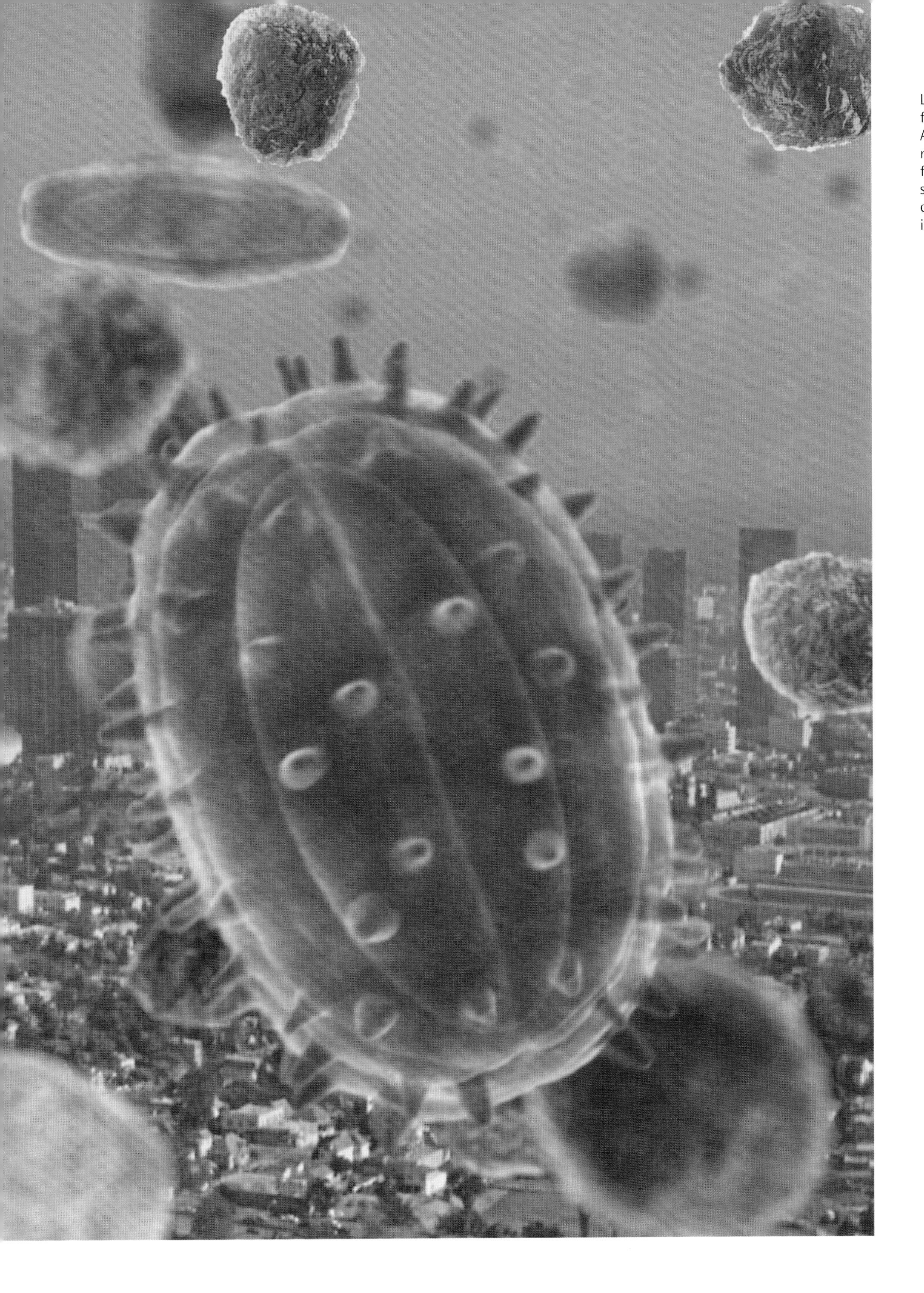

Left: Not even the freshest air is truly clean. A lungful will contain millions of particles, from industrial waste to spores of fungi, as this computer-generated image shows.

tornado alley

When it rejoins the jet stream miles above us, our notional parcel of air is now half-way to the pole. It is just about to become part of one of the most infamous forces of wind on the planet. Its name comes from the Latin for "to turn," *tornare*. It is the tornado.

The stories associated with tornadoes are some of the most bizarre in the records of meteorology, and the author Lyall Watson has gathered together a veritable feast of strange tales of survival and disaster, most of which make the adventures of Dorothy in *The Wizard of Oz* seem plausible. There was the house in Oklahoma that was picked up, twisted round and set down facing in the opposite direction; the house that rose so gently that its owner fell 30 feet (10 meters) to the ground when he stepped out to see what was going on; a train taken from its tracks and dropped in a ditch; wells and rivers sucked dry; a mattress sucked through a window without waking the child who slept on it; horses lifted up and left astride a barn; milkmaids abandoned with just a bucket as their cow disappeared; and a herd of cattle that was seen floating away like a flock of birds. But the stories of absurdity hide the deadly nature of a tornado. Victims are beheaded, torn to shreds, coated in mud and their wounds filled with splinters of debris.

Scientists have begun to use the remarkable ability of a tornado to carry objects over long distances to help them understand the physics of the phenomenon itself. A tornado debris project currently underway is mapping the fallout from tornadoes, tracing its origins and working out the nature of the forces that were needed to transport it. Today's science will yield valuable results for both meteorologists and designers of buildings, but there is a less edifying history to tornado studies. In 1842 one Elias Loomis decided to test a theory as to why chickens vacuumed up by tornadoes were often found

Right: The funnel of a tornado is marked by the dust and debris that it draws up.

intact on the ground, except for being totally devoid of feathers as if perfectly plucked. Loomis believed that in the very low pressure inside the whirlwind the air sacs inside the quills literally exploded, thus removing the feathers cleanly. So he arranged a simple experiment, in which he fired a cannon straight up into the sky, using a chicken instead of a cannonball, to find out what wind speed was needed to do the plucking. The result was a cloud of feathers that rose into the air and was scattered on the wind, while the chicken totally disintegrated and the pieces were never found.

The real impetus for modern tornado research came after April 1974, following the worst tornado outbreak ever recorded in the history of the USA—or even of the world. Over a period of just sixteen hours 148 twisters touched down across thirteen states, leaving 315 dead, 5,484 injured and entire townships obliterated. There were six F5 tornadoes—the strongest on the scale—and one was

Above: The vast thunderhead that is the real power behind a tornado.

5 miles (8 kilometers) wide. The meteorological factors that came together to create the "super outbreak," as it is now dubbed, were typical of the region. The stretch of the United States which runs from Kansas through Oklahoma to Missouri is known as "tornado alley" with good reason. The Great Plains of the USA are heated by the strong summer sun. This heat expands the air above the ground, causing it to become lighter and to rise, thus sucking more air in to replace it. But meanwhile, out over the Gulf of Mexico, there exists a vast source of very moist, warm air, and this can be dragged into the suction process. That's what was happening in April 1974. But there was also a flow of colder air coming in from the west, undercutting the tongue of warm air from the gulf as it moved further north towards the plains. As the warm, moist air rose over the cold, dry, dense air, it got pushed further up into the cold atmosphere where its water vapor began to condense, forming clouds and eventually thunderstorms. and then the jet stream began to get involved. There was a strong jet stream flowing directly over the collision area, pushing colder air in from the west and therefore speeding its movement. This actually intensified the creation of thunderstorms by causing even more rapid uplift of the warm, moist air and thus increased the likelihood of severe weather.

Normal thunderstorms, despite being powerful enough to produce dramatic lightning shows, cannot form the deadly cocktail of grapefruit-sized hail, torrential floods and, above all, tornadoes. Fortunately only a small number of thunderstorms go on to become true weather beasts, because for this to happen something else is required: they need to last for hours to build up the potential to spin out tornadoes—and to do that they need to be capped.

To the south-west of the Great Plains lie the desert regions of the USA. These are at a much higher elevation and, to complicate things, warm, dry air flows off them and moves across Tornado Alley at a height of about 3000–6000 feet (1–2 kilometers) above the ground. This layer of warm, dry air is known as a temperature inversion, and the warm, moist air from the gulf that is rising below it is not warm enough to rise any further—it is capped, so at first no small thunderstorms are produced. But as more and more warm, moist air flows in and rises up, this inversion layer begins to act like the lid of a pressure cooker, with more energy building up beneath it but nothing allowed to boil over. The cap suppresses storm formation until, by later in the day after the clear skies have allowed the sun to heat surface air even more, it is so moist and warm that, when it rises, it is warmer than even the warm air of the cap. Then it breaks through. Just as with a pressure cooker, the release is all the more violent for being delayed, and the resulting thunderstorm grows much, much larger. The warm air can then rise to a colossal height, up to the height of the jet stream itself.

Now the jet stream gets directly involved. Air flowing swiftly through it takes with it some of the warm, moist air that has risen up. This means that more warm air has to be sucked up from below, replenishing and adding to the moisture and doubly intensifying the updraft, setting the stage for the final violent phase. These were the conditions that prevailed in April 1974.

A severe thunderstorm can last for hours. As rain condenses out and begins to fall from the towering clouds it cools the air, making it heavier and prone to sinking, and creating a downdraft (see "Thunderstorms" on pp 92–3). As this cooler air hits the ground it then spreads outwards, pushing more warm air up into the cloud and thereby enhancing the updraft. This feeds the storm with heat and also with energy and so the process just keeps going. Thunderstorms such as these now go on to

Above: The spiraling air below a thunderstorm begins to rush upwards to form the funnel of a twister.

produce the most dramatic phenomena of the weather.

For a thunderstorm to create a tornado the storm must rotate. It is not clear exactly how this begins, but some scientists favor the following explanation. As the thundercloud grows higher, the cloud experiences strong winds going in different directions within it. This "wind shear" causes turbulence, with some of the cloud rolling over itself, and this can create a horizontal spinning vortex of air near the surface—a bit like a pipe rolling over and over in the same place. What happens next is extraordinary. A strong updraft of rising air can push the middle of the vortex up into the cloud, bending it into the shape of a hoop and leaving the two ends of the spinning vortex touching the ground. This means that two main vertical vortices are formed, one to the north and one to the south, with the northern end of the vortex rotating clockwise and the southern end rotating anticlockwise. Then the falling rain in the storm is pushed towards the north-east by the strong, high-level winds, where it meets the

clockwise vortex of air and sets the downdraft spinning. Once that has happened, the updraft starts to form in the anticlockwise vortex at the southern end. This is where the tornado is born. As air rushes into this upward spiral it expands and cools, and condensation forms a visible cloud—the funnel. The cooling and condensation accelerate, the funnel descends towards the Earth's surface and the terror begins.

Upon reaching the ground, the tornado usually picks up dirt and debris that make it appear dark and ominous. While air along the outside of the tornado is spiraling upwards, within the core of the tornado the air is descending towards the ground. It does this because air is needed to replace the rising air in the updraft. As the air descends it warms, causing the cloud droplets to evaporate, leaving the core free of clouds.

One severe thunderstorm and tornado is enough, but what made 1974 so horrific was that chains of them sprang up, one after the other in line after line. As the cold air was swept in from the west it would have moved in surges, which caused the warm air confronting it to move in waves, and the rising waves would have triggered a sequence of uplifts and growth of storms. Then the thunderstorms actually supported each other in their growth, the downdraft of cool air from one thunderstorm reinforcing the updraft of a neighboring storm by pushing warm, moist air upwards. Normally these "supercell" storms develop in a single line, known as a squall line, but in April 1974 there were three of them. With so many thunderheads developing, the lines of storms sheltered each other from high-level winds that could have fragmented their peaks before they had a chance to grow fully. It was a truly awesome combination of natural forces, and one that tornado alley had little chance to do anything about.

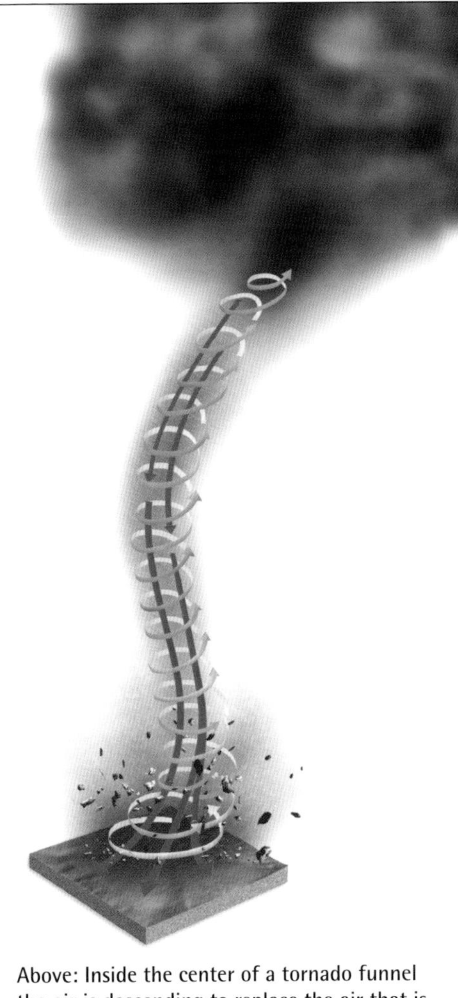

Above: Inside the center of a tornado funnel the air is descending to replace the air that is spiraling up the outside, forming the violent "twister" as it rises.

During the course of the April 1974 outbreak, ninety-seven tornadoes were declared to have been violent, and 600 million dollars' worth of damage was done. But the sixteen hours also dispelled one of the age-old myths about tornadoes: that they do not travel up and down steep hills. In fact, a tornado that hit Guin, in Alabama, stayed firmly attached to the ground as it climbed inexorably over the 1,600-foot (500-meter) Monte Sano mountain; and on the same day the Blue Ridge tornado crossed a 3,000-foot (1,000-meter) ridge, traveled down the canyon on the other side and climbed another 3,000-feet (1,000 meters) before dying out.

inside a tornado

Few people have seen inside the vortex of a twister and survived, but Bill Keller from Kansas lived to describe the cloud that passed over him in June 1928: "A screaming, hissing sound came directly from the end of the funnel, and when I looked up, I saw to my astonishment right into the very heart of the tornado....It was brilliantly lit with the constant flashes of lightning zigzagging from side to side....Around the rim of the great vortex, small tornadoes were constantly breaking away and writhing their way around the funnel." Apart from the violent destruction wrought by the fast-moving winds, the inside of the funnel contains a very low air pressure—the equivalent of the pressure difference between ground level and an altitude of 4,900 feet (1,500 meters)—which means that the suction is phenomenal. And Bill Keller's description of whirlwinds within the whirlwind has been borne out by recent research. Professor Josh Wurman of the University of Oklahoma has identified a quite unexpected shape to the inside of a tornado vortex. On the first Doppler radar images it appeared to be square, but then his analysis revealed a complex pattern of vortices rotating around the main vortex, analogous to a ring of ball bearings, which he was able to measure moving around the tornado for over eight minutes. These smaller "suction" vortices reach phenomenal wind speeds, as each rotates at perhaps 125 miles (200 kilometers) per hour around a twister that is itself already rotating at that speed. The damage caused by a multiple-vortex tornado is of the worst kind.

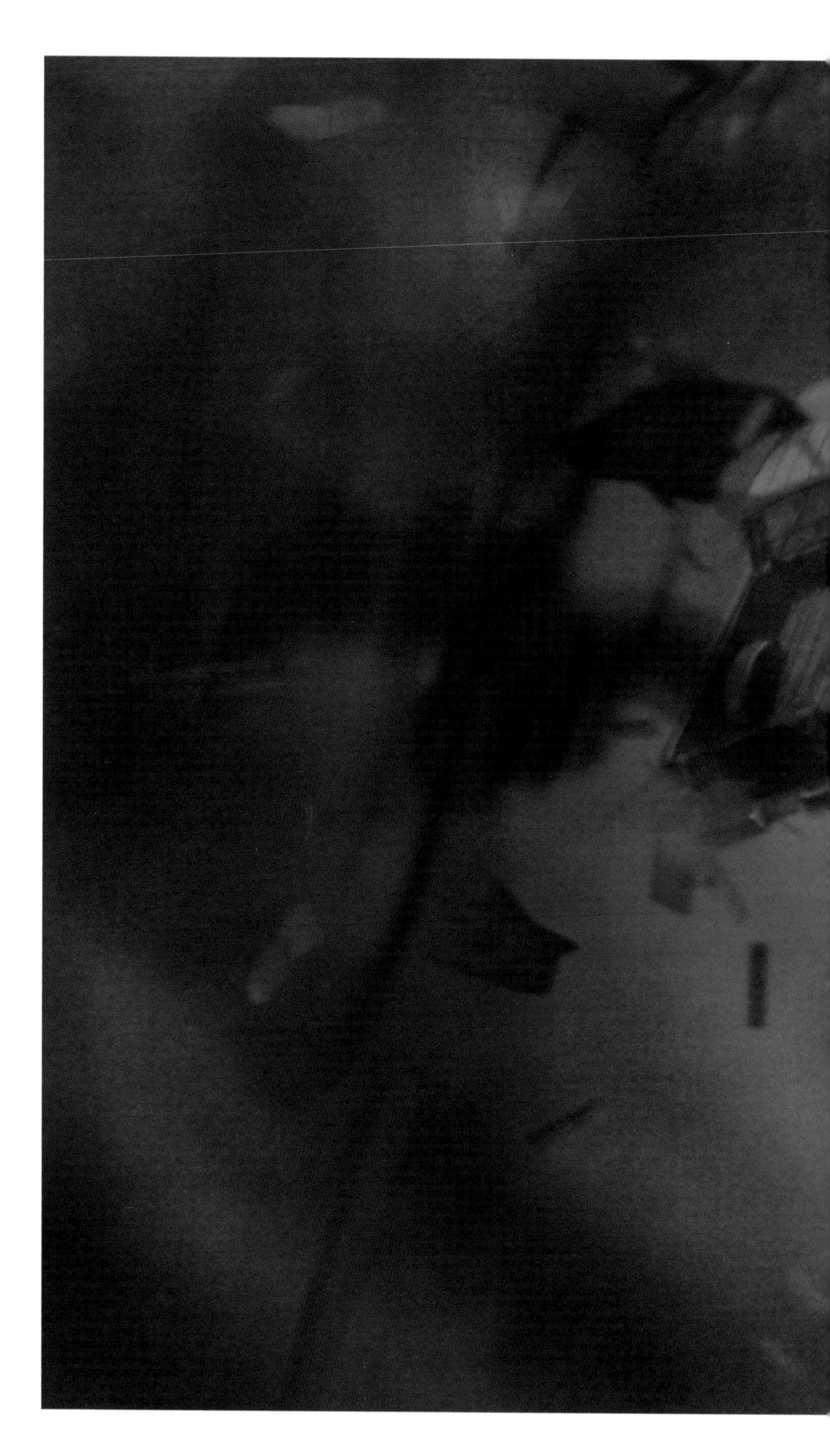

Right: A computer-generated image of the inside of a twister: a maelstrom of debris, a tight spiral of cloud roaring upwards with lightning playing around the inside.

TORNADO WARNING

Today, huge arrays of technology are used to study and predict the movement of tornadoes, because the richest country in the world, the USA, also just happens to be the country which reports more tornadoes than anywhere else on the globe. The tornado that swept through Oklahoma City on May 3, 1999 destroyed one thousand buildings and caused damage running to billions of dollars, carving a path 30 miles (50 kilometers) long and up to 580 yards (half a kilometer) wide. Forty people died in tornadoes across Oklahoma that day. While such a beast is unlikely ever to be stopped, early warning of its arrival, and designing property to withstand its force, can now save money and lives. Scientists have a scale of measurement for tornadoes, created by University of Chicago meteorologist, Theodore Fujita, and named after him as the F-scale, which runs from F0 to F6. He gave each grade a descriptive word: F0 was "light," then came "moderate," "considerable," "severe," "devastating," and finally the "incredible" F6. The Oklahoma twister was a "rare F5," the most extreme that he defined, and it was part of a "swarm" of tornadoes that included an F6—what Fujita originally listed as "inconceivable"—with airspeeds in the column calculated at over 300 miles (500 kilometers) an hour. Calculations are all that can be expected, because no instrument could reliably withstand the maelstrom that is the funnel of the twister, as it sweeps up anything and everything in its path, carries the debris onwards or upwards for as long as it sees fit and then spits it out, up, down or sideways.

FUJITA SCALE FOR TORNADOES

Grade	Wind speed	Damage
F0	40–71 mph	Light damage
F1	72–112 mph	Moderate damage (roof tiles gone)
F2	113–156 mph	Considerable damage (windows burst and roofs lifted)
F3	157–206 mph	Severe damage (outer walls collapse)
F4	207–260 mph	Devastating damage (house destroyed)
F5	261–316 mph	Incredible damage (structure and rubble blown away)
F6	Over 318 mph	Inconceivable

Back at the time of the "super outbreak" of 1974, the weather-service forecasters could only see green blobs on their radar screens, which told them almost nothing reliable, and they had to wait for a visual sighting before issuing a tornado warning. Today they can view the evolution of a thunderstorm in detail and issue warnings with confidence before the twister has even begun to form. This phenomenal advance stems principally from the development of the technique called Doppler radar. A conventional radar signal is able to pick up an echo from droplets of water raining out of a cloud, and so meteorologists had learned to identify telltale patterns of precipitation inside a thunderstorm which might indicate that a tornado was about to form. Perhaps the most well known of these is the characteristic shape of a hook reaching out from the center of the storm. The hook was where the twister would always form. But twisters would often form with no "hook echo" present, and on other occasions the hook would show up after the tornado had touched the ground. Doppler radar, on the other hand, can do more: it is effectively the same as the device the police use to detect speeding motorists and can measure the speed at which rain is moving towards or away from the detector. As the rain in a developing tornado is carried around with the wind, this technique means the strength of the vortex can be measured precisely. Two radar instruments facing the storm at different angles are needed to build up the full picture, and today the scientific storm-chasers, dominated by the team from the University of Oklahoma, take with them DOW (Doppler on wheels) units as they track and observe the tornadoes they study. Their work has now revealed in remarkable detail what actually goes on inside the vortex.

Tornado science has come a long way in the last thirty years. Doppler radar is only one technique being applied to understanding these remarkable storms. There are also approaches that rely on seismic tremors that radiate from a tornado, and in addition tornadoes give out acoustic signatures, sounds that are below our hearing range but which instruments can detect up to 125 miles (200 kilometers) away. Some scientists believe that in the future not only will they be able to predict precisely the emergence of tornadoes, but they will also be able to destroy them. Some even believe that we could eventually harness the energy they give off!

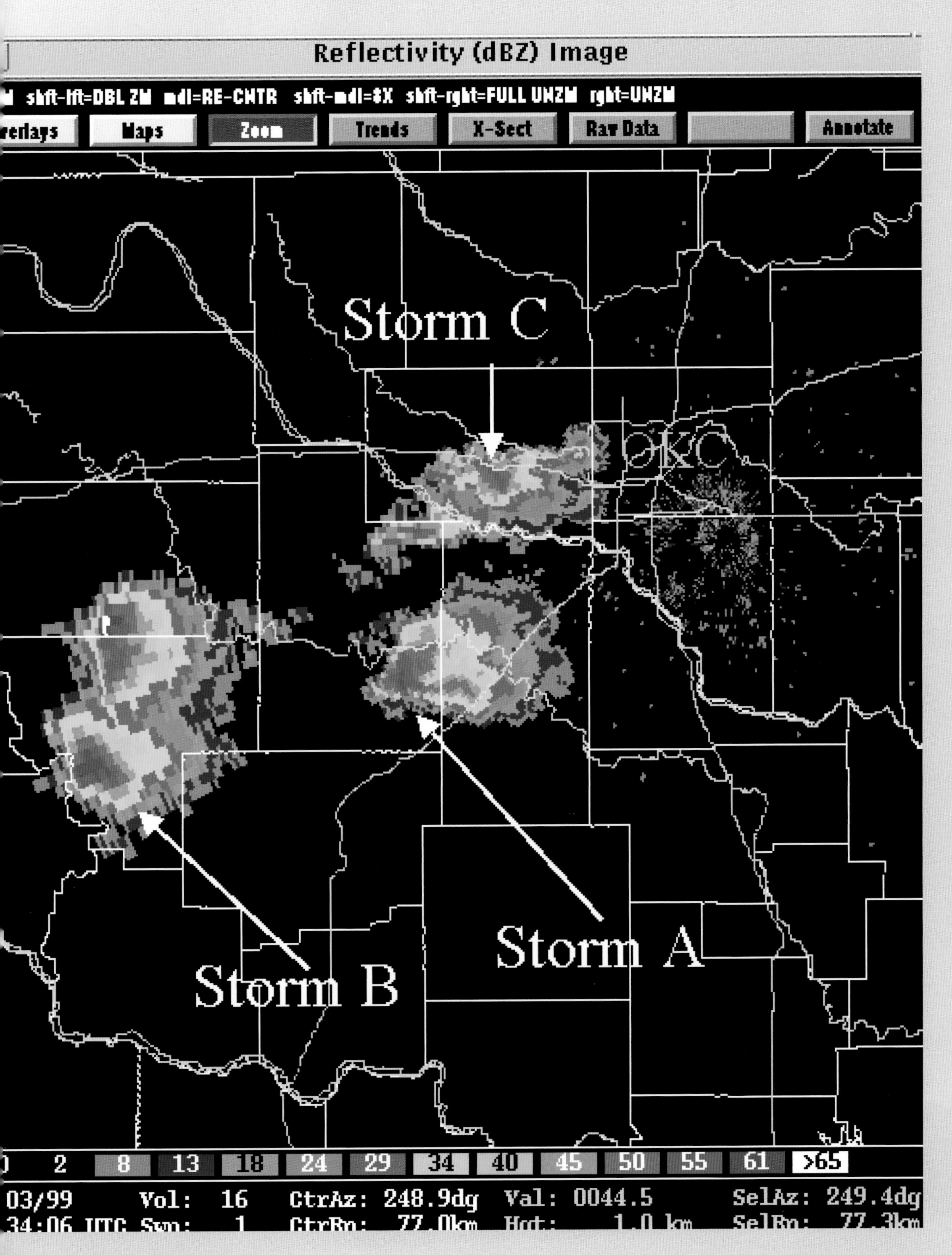

Left: Doppler radar images of storms likely to form tornadoes over Oklahoma.

Above: The storm that devastated the Fastnet Race off Ireland in 1979 was born as far away as the North American plains.

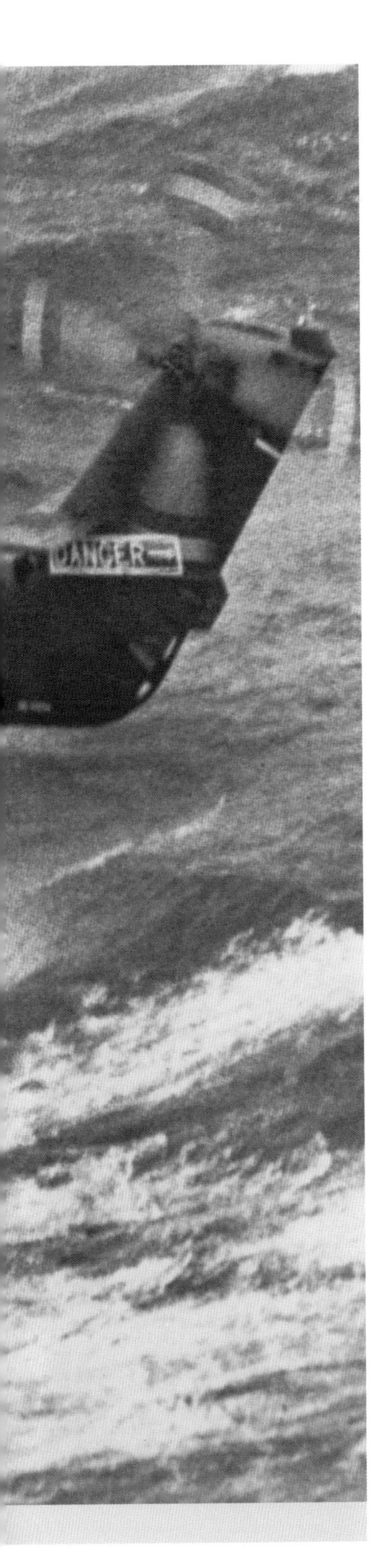

north atlantic winds

The violent thunderstorms of the Great Plains routinely blow themselves out when the supply of warm, moist air at their base runs low, the updraft fades, the supercell rotation slows and their energy is dissipated in wind, rain, lightning and hail. But occasionally a storm will continue to feed itself, as it sweeps across the open land, heading north-east towards the polar cell, for the final part of the global journey of the winds towards the pole. On August 9, 1979 just such a monster developed over Minnesota and moved inexorably across the USA to the eastern seaboard and the northern Atlantic Ocean. It was not a tornado, but it was a huge thunderstorm nonetheless. From torrential rain over Minneapolis on Thursday, it was blowing tollbooths off the New Jersey turnpike and downing trees in New York's Central Park on Friday. By Saturday it was over Halifax, Nova Scotia, and on Sunday it left the coast for the ocean—about a day after the flotilla of 303 yachts competing in the annual Fastnet Race set out from the Isle of Wight on the south coast of England to sail to the Fastnet Rock off the tip of southern Ireland.

A massive high-pressure system to the south and a deep low over Iceland meant that the storm was blown across the Atlantic towards the Bay of Biscay, where it would have caused the competing sailors little trouble. But it never arrived there. Instead, the Icelandic low pressure stalled and the storm from America moved on round it, traveling at over 600 miles (1,000 kilometers) a day, to a point where southerly winds pushed it north towards the cold polar front that marked the boundary with the polar cell. It met cold air sweeping down from Iceland and intensified as a result, reaching force 10 as it hit the center of the race. The deep low pressure of the storm, the fourth lowest barometric reading for the Atlantic summer since the year 1900, caused a vast mound of water to rise up beneath it as it sped across the ocean. That in itself set up a huge rolling wave motion. The 60-mile- (100-kilometer-) an-hour winds whipped up even more waves on top, including rogue waves—individual walls of water that move across the path of the main wave pattern—which, with the wind speeds that occurred, reached twice the height of the mast of an ocean racer. Twenty-four hours later fifteen people had died, five yachts had sunk and 136 people had had to be rescued. The storm was very compact, and moved so fast that forecasters were barely able to register its potential threat as it traveled through their charts, but it was one of the most violent passages of air across the Ferrel cell that could be imagined.

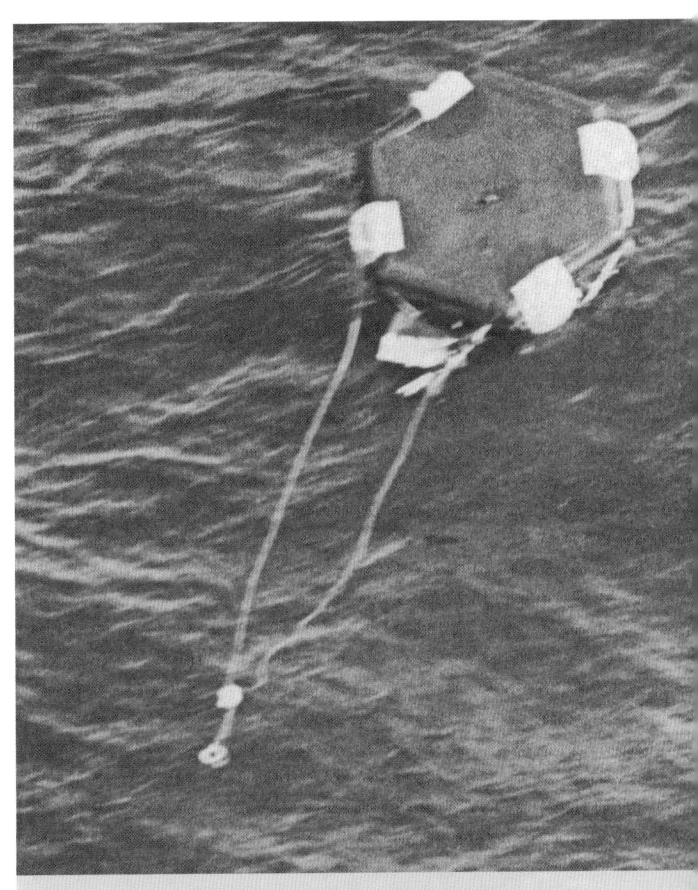

Above: Dragging its anchor, an upturned life-raft from one of the Fastnet yachts drifts off Land's End in Cornwall.

solar winds

Eventually, the airflow from the Equator, via hurricanes, jet streams, ill winds, tornadoes and ocean storms, reaches the boundary with the polar cell, where some of it now rises to be carried on the high-level flow towards the North Pole itself. But up here it flows beneath another kind of wind altogether—one that comes from 93 million miles (150 million kilometers) away, and which becomes visible only in the dark nights of the polar winter when it creates one of the most awe-inspiring phenomena it is possible to see on the planet. It is the solar wind, and the magical experience that it ignites is the aurora borealis, the northern lights.

Constantly flowing out from the polar regions of the sun itself comes a stream of charged particles—protons, electrons, hydrogen and helium—that rush away across space. This "solar wind" reaches far beyond the corona of the sun, past the inner planets, including Earth, past Jupiter and Saturn, and out beyond Pluto, to the very edges of the solar system. It flows alongside a deadly stream of radiation of ultraviolet, gamma rays and X-rays, and would mean the end of life on Earth were it not for the remarkable magnetic shield that our planet has thrown up around us. Discovered only as recently as the 1960s are force fields that wrap around the world from a level of around 60 miles (100 kilometers) or more and upwards into

Above: The green glow of the aurora borealis over a field of ice, with the moon just setting.

Above: The aurora australis reaches out above the south pole. Captured on film by the crew of a space shuttle.

the sky. Known as the Van Allen belts, these make up the Earth's magnetosphere, lines of its magnetic field that loop out from each pole, curve round the planet and turn back in to the opposite pole. High above our weather this magnetosphere deflects the deadly wind from the sun, but at either pole the charged solar particles are channeled down in waves towards the surface of the planet and reveal themselves as the aurora. The solar wind travels at 248 miles (400 kilometers) per second. Four days after leaving the sun its particles hit us, and vast amounts of our atmosphere are blasted away into space in the ensuing battle. In space it is a frighteningly dramatic scene, but on the ground in the Arctic the effect is somewhat different. The solar winds bring perhaps the greatest show on Earth.

For the sun is not a constant thing. It is now known that the intertwining magnetic fields that curve out from its own poles, and which twist like a braid around it, become more knotted and stretched until there is a moment of rapid rearrangement back into simpler lines of force. In places the field lines can curve back in, literally holding light back from emerging from the surface and creating the dark patches we can see from Earth as sun spots. This slow, tortuous build-up occurs over a period of eleven years, ending with what is known as a solar maximum. This is when the violent outbursts that occur and the discharge of particles from the surface are at their most frequent, have their greatest reach and are at their most deadly. As the solar maximum draws near, the turmoil of solar storms on the sun's surface dramatically increases the occurrence of coronal mass ejections—huge loops of ionized gas, or plasma, that rush out across space energizing the solar wind. The result for us is more violent geomagnetic storms that rage in the magnetosphere of the Earth as our magnetic shield battles to protect us. The interaction with the Earth's magnetic field creates a backlash of particles that are funneled into two rings around each of the Earth's poles, known as the auroral ovals. These are where the continuous stream of charged particles creates the aurora borealis in the north and the aurora australis in the south; and the additional charges that emerge from the geomagnetic storms dramatically increase the

Left: A computer-generated image showing the effect of the charged particles of the solar wind on the Earth's upper atmosphere at 248 miles (400 kilometers) a second.

Below: The shape of the Earth's magnetosphere is distorted as it shields us from the effects of the solar wind.

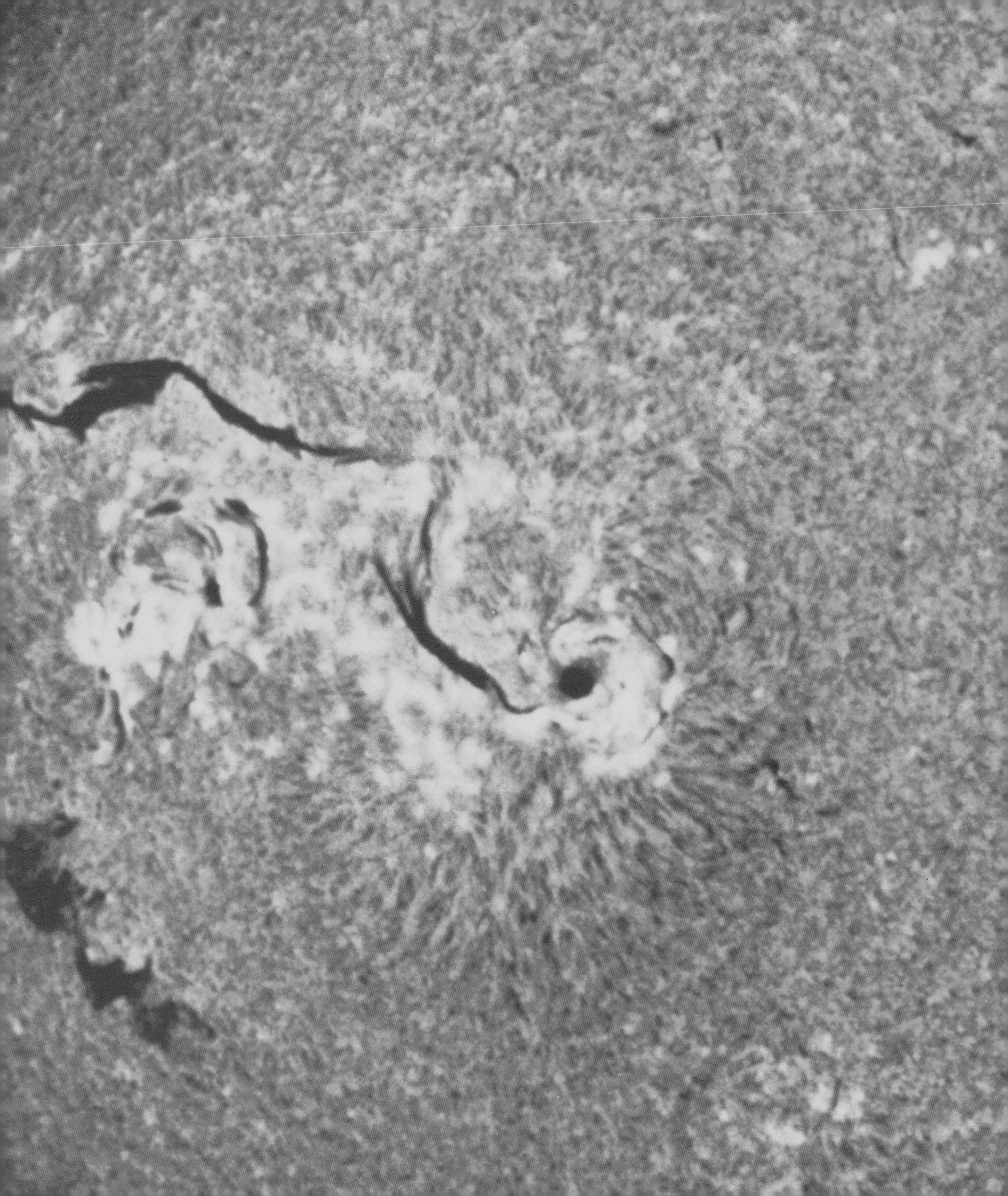

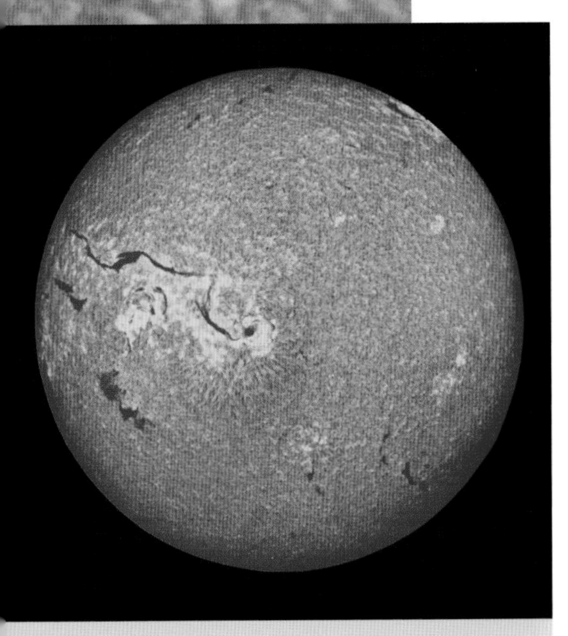

Above: Periods of sun-spot activity seem to coincide with climatic changes on Earth.
Left: Sunspots are darker, cooler areas of the sun's surface, created by magnetic field-lines that arc out and back into the heart of the star.

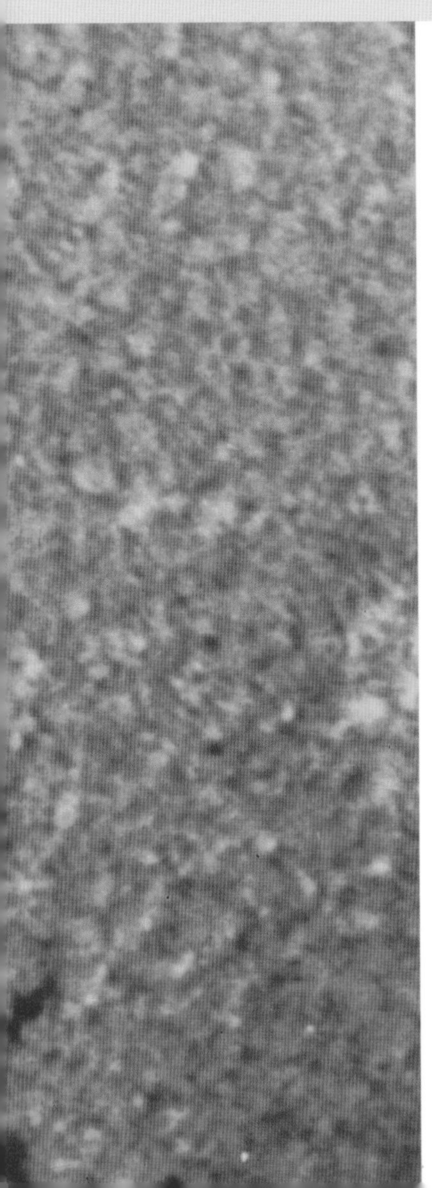

light shows of the northern and southern lights.

The Inuit people of Labrador believed that the aurorae were torches carried by spirits who were showing those who had died violently the way to heaven, and that they could see the spirits in among the auroral lights as they happily played a game of ball with the skull of a walrus. They also believed that a whistling or cracking sound could be heard from the aurora. Some explained this as the sound of spirits running as they played on the frosty snow in the heavens, while others knew it as the whispering voices of the spirits trying to talk to those left behind on Earth. Today there are still many unanswered scientific questions about the aurora and the interaction of the solar wind and our planet. The sound that many report they can hear during displays of the northern lights has defied explanation. And the increasingly apparent link between the activity of the solar wind and our own weather on Earth is also still a puzzle.

For many years now it has been realized that sun-spot activity appears to coincide with changes in our weather and climate. At the grand scale, it is known that the eleven-year cycle was not always constant. There was a period called the "Maunder minimum" between 1645 and 1715 when sun spots were almost non-existent, and it seems that this coincided with a particularly cold stage in our history known as the Little Ice Age, when frost fairs were the norm and the River Thames froze in the London winter. But more recent observations suggest that periodic droughts on the Great Plains of the USA seem to correlate with a twenty-two-year cycle of solar activity, and the eleven-year cycle has a relationship with winter warming in the northern hemisphere and with a pattern of stratospheric winds reversing in the tropics. No explanation has been agreed, but there are also tantalizing suggestions that, on a much shorter timescale, the solar wind could be working to intensify polar storms. The theory is that the high-energy particles from solar flares reach down to touch the troposphere, the part of the atmosphere that is closest to the surface of the Earth, where they help produce condensation nucleii, the small particles on which water vapor condenses to form clouds. The increase in cloud cover would therefore mean more heat was trapped inside the storm, and so the energy of the storm would be greater and the storm more intense. It is a plausible theory, and, even if scientists do not yet know the correct mechanism, they are beginning to accept that the forces that drive our weather do not stop at the boundary between Earth and space, but reach back as far as the source of our very existence: the sun.

journey's end

Finally the air high aloft, passing through storms, moving from the warmer south, reaches the frozen North Pole itself. There our journey with the winds has reached an end, for the air cools over the pole and inevitably sinks exhausted. Now at last the cold polar air can begin its own journey back towards the Equator, across the cells of global circulation, gliding first across the Arctic Circle and pushing south behind the polar front. There it rises to speed on in the jet streams high above the westerlies of the mid-latitudes, and sinks down again to create the steady easterly flow of the trade winds. Finally it comes to rest in the calm of the doldrums, where it waits to begin one more round of the great global circulation of the weather.

Water dominates our world: the sea surrounds us, rivers scour the land and there is also an ocean above us. Every drop of water on Earth undergoes an extraordinary journey through time and space, constantly changing its form. It can rise from the oceans, flow through the sky, fall upon the Earth, freeze for millennia—creating the clouds, rain and snow at the heart of all our weather.

chapter three

wet

wet but beautiful

Settled high in the Khasi-Jaintia hills of the Meghalaya region of India, lying at longitude 91 degrees 43 minutes east and latitude 25 degrees 17 minutes north, Cherrapunji is a small town of immense significance. Behind it rise the limestone cliffs of the Shillong plateau, lined with lush greenery and muddy clay soil and often topped with dense, white clouds. Deep chasms are carved into the limestone, bearing rivulets and spectacular waterfalls that rush down through the town and on towards the distant flood plains of Bangladesh that can be viewed through a blue haze to the south. Far to the north are Mount Everest and the peaks of the Himalayas, the remote state of Bhutan and, beyond it, the fabled land of Tibet. But what makes Cherrapunji so very special is that it marks the furthest reach of the Indian monsoon, during which the entire subcontinent is engulfed by its annual rain. It is a place whose population waits in tense anticipation as the thunderous storms of the summer sweep across the Bay of Bengal and draw near. When at last they strike Cherrapunji, it rains like nowhere else on the planet. For many years this remarkable place has held the record for the greatest amount of rain ever to fall in one place in one year. It has a reputation as the wettest place on Earth.

Above: A waterfall cascading off the Shillong plateau towards Cherrapunji in north-east India.
Right: Almost constantly topped with cloud, Cherrapunji held the record as the world's wettest place for many years .

water world

If the wind is the engine of the weather and the heat from the sun is the energy that creates the ignition, water is the fuel. We live on a water planet. The seas dominate our world, but there is also an ocean around us and an ocean above us. Water is in all our cells. It is at the heart of all living things and it is at the center of all our weather. Although we use it and consume it at an ever-increasing rate, the amount of water on the planet is largely unchanging, and the same water molecules have remained in existence for hundreds of millions of years. What happens on Earth is that the same water is used and reused. It rises from the oceans, rests in and flows through the sky, falls upon the earth around us and finds its way back inevitably to the sea. Indeed the main activity of the planet might be regarded as being the transport of water, and water's journey through both time and space is as surprising and profound as are its effects upon us.

The global water cycle touches us in many ways. Technically known as the hydrological cycle, it consists of six basic processes: evaporation, condensation, precipitation, infiltration, runoff and transpiration. It goes like this. The sun's rays heat the water at the surface of the Earth's oceans and lakes. The water molecules become so energized that some of them turn into vapor—they *evaporate*, and rise up into the air as a result of convection (warm air rises and cool air falls). But as the water vapor rises it loses energy, cools and *condenses* into the liquid state, creating clouds of water droplets. The droplets of water at some point become too heavy to remain in the air, and begin to fall as *precipitation*, in the form of rain, snow or hail, on to the surface of the Earth. There the water *infiltrates* the soil and rock, seeping into pores, cracks and cavities where it can rest for millions of years as groundwater, or be drawn up by the roots of plants as they extract their mineral nutrients from the soil. Alternatively, depending on how hard the ground is, or how saturated it already is with water, the rainfall will flow in streams and rivers as *runoff* into lakes or the oceans, passing through many living things along the way—including us as we drink it up and allow it to drain away. Finally the cycle is completed as the sun heats the water again, which begins to *evaporate* back up into the air. There it joins other vapor that has been *transpired* into the atmosphere by plants as they draw their mineral nutrients up their stems, displacing water as they go.

The quantities involved are staggering. The amount of water in the atmosphere at any one time is almost 3,000 cubic miles (13,000 cubic kilometers), a quantity that if poured on to the USA would leave the entire population wading to a depth of 4 feet (1.3 meters), or in the UK would drown those living there in water up to a depth of 175 feet (53 meters). The amount that falls on the Earth each year is 140,000 cubic miles (578,000 cubic kilometers), which works out at over 4 billion gallons (18.3 billion liters) per second—about 5 pints (3 liters) per second per person on the planet (although, sadly, it is not evenly distributed, so some endure drought while others suffer flood). But because the process is a cycle, it means that 4 billion gallons are being evaporated every second, and the amount of energy required to do that—every second, remember—would power the whole of the UK for two weeks.

WATER POWER

There is a chilling statistic which reveals the awesome power of water. It is the amount of erosion that the water cycle creates as rain, gales and hurricanes scour the surface of the Earth. Twenty billion tonnes of eroded rock and soil are transported from the continents to the oceans every year, and the energy locked up in this global stream-flow is the equivalent of four million atomic bombs being exploded every year. Water shapes and sculpts our world—sometimes over years, sometimes overnight. To be impressed by the extraordinary power of water to shape the world it is only necessary to make the journey to the Grand Canyon early in the morning or late in the afternoon, when the sun picks out the shadows of the gullies, towering buttes, deep valleys and wide mesas that have been carved within its gorge. Two hundred and twenty miles (350 kilometers) long, 20 miles (30 kilometers) wide and over 1 mile (1.5 kilometers) deep at its deepest point, this vast gash in the surface of the planet has been cut by the slow and steady action of the water of the Colorado river, over a period of six million years—and every drop of that water came from the sky.

While the power of a river to carve rock slowly is immense, more frightening is the phenomenon of the flash flood, a flood which rises rapidly with no advance warning. And it is in canyons such as the one carved by the Colorado river that the flash flood is at its most deadly. On July 31, 1976 two thousand people, mostly campers and visitors to the Big Thompson canyon enjoying a holiday weekend on the slopes of the Rocky Mountains in Colorado, experienced one of the worst in recent American history. It was a warm afternoon, with a light easterly wind playing against the warm eastern foothills of the mountains—much like any other summer's day. During the course of the afternoon small cumulus clouds began to build up with their distinctive flat bases and dome-shaped tops, and gently became pushed up the sides of the mountain range as the air warmed. Some of the clouds were seen to have flattened tops, suggesting the presence of a temperature inversion, where warm air lay above cold, preventing any air from easily rising further and encouraging the clouds to build up against the mountains instead of rising up and being blown away. Towards the end of the afternoon some of the cumulus clouds pushed their way through the inversion and began to build more rapidly, fueled by warm, moist air blowing in from the east on strong, low-level winds, until they formed huge, multicell thunderstorms that rose as high as 65,000 feet (20 kilometers) into the air. Heavy rain began to pour. High above, the high-level southerly winds were unusually weak and, instead of pushing the storm away to follow its normal track to the north, they failed to move it at all. For several hours the giant cumulonimbus clouds stalled in one place, right over the narrow walls of Big Thompson Canyon. The rain became torrential—12 inches (30 centimeters) fell between 6.30 and 10.30 that night, with more than half the region's annual rainfall coming down in a single hour! The walls of the canyon were solid rock, with nothing to absorb the water. All of it flowed into the river, which filled to capacity and burst its banks at the narrowest part of the gorge. The road was flooded, battered by the turbulent water, and collapsed. The campsite, mobile homes, cars, tents and cabins were all caught up in the flow, and the debris created a huge natural dam in the river. Even more pressure built up behind the blockage until the pile of rubbish gave way and a wall of water tore down the valley at a speed of 50 miles (80 kilometers) an hour. Over 135 people died.

It was one of the worst natural disasters of the post-war years for the USA, but nowhere near the last of its kind. Elsewhere in the world floods and mudslides due to torrential rain have killed many more just since the mid 1990s: 63 in Sudan, 71 in Spain, 156 in Tibet, over 600 in Mexico in 1999 and—the worst of its kind in recent history—up to 50,000 in Venezuela in that same year.

Left: The Colorado river is still actively carving new valleys along the length of the Grand Canyon.

Below: Six million years of flowing water has resulted in one of the greatest natural sculptures in the world: the Grand Canyon in the western USA.

raindrops

This thing that does so much damage is vital to our existence and is itself beautiful: a raindrop. The air above us is not the clear, empty space that we might feel it to be. Even on a crisp spring morning, when a view from the South Downs of England might afford a glimpse of the French coast on the other side of the English Channel, even on such a perfect clear day the air will be full of things that we can only begin to guess at.

Rain begins in a world that is infinitesimally small. At the heart of every raindrop is a condensation nucleus. Measuring an invisible ten-thousandth of a millimeter across, it is a substance that is hygroscopic, or "water-absorbing." So a molecule of sea salt, or a compound of ammonium sulphate or nitrate, or a microscopic fragment of organic matter caught up by the wind, or pollen, bacteria or dust—each can have the magical property of drawing to its surface the water vapor that exists in the air; over the sea, a quart of air might contain a million of them, over land it might be five million. In warm air the vapor molecules are so active that they may strike the surface of the condensation nuclei and simply bounce away; but in cooler air the vapor molecules are less frenetic and will stick. And when some billions of these molecules have stuck to a nucleus, a droplet will have formed—a cloud droplet that is by now a more substantial size, reaching perhaps two-hundredths of a millimeter in diameter. Cloud droplets are so small that the tiniest of upward air currents will keep them aloft, and countless billions of them are what you see in the form of fluffy, white blobs of cotton-wool clouds that scud across a summer sky. The droplets would grow with more water vapor condensing on to them, but it would take days for them to reach a size sufficient for them to start falling. Instead, what happens is that as

Above: The shape of raindrops depends on their size: as they fall, air resistance deforms larger drops from a

they swirl around in the air high above your head they collide and coalesce together to form larger and larger droplets until, by the time about a million cloud droplets have combined, the updrafts of air can no longer support their weight and they fall. In this microworld the falling droplets are still resisted by the air around

sphere to a flat pancake shape.

them. The larger the drop, the greater the resistance of the air, but the greater also is the force of gravity that pulls it down. Eventually gravity and resistance balance out so that droplets fall at a constant speed, with larger ones falling faster than smaller ones. As they fall the larger drops overtake smaller ones, striking them and coalescing with them, thus getting bigger and faster—by now reaching .07 inches (2 millimeters). They have become rain.

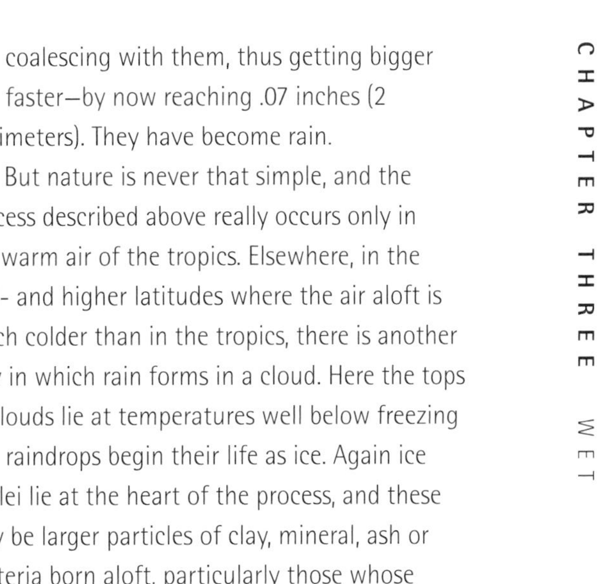

But nature is never that simple, and the process described above really occurs only in the warm air of the tropics. Elsewhere, in the mid- and higher latitudes where the air aloft is much colder than in the tropics, there is another way in which rain forms in a cloud. Here the tops of clouds lie at temperatures well below freezing and raindrops begin their life as ice. Again ice nuclei lie at the heart of the process, and these may be larger particles of clay, mineral, ash or bacteria born aloft, particularly those whose surface shape encourages the angular structure of an ice crystal to grow upon it (see Chapter Four). Instead of condensing as liquid, the water vapor forms an ice crystal, which itself will grow, eventually combining with other crystals to become a snowflake as it falls. Depending on the air temperature below, this may reach the ground as snow or melt to become rain. Indeed, most of the rain outside the tropics begins its life as snow.

However they begin, as larger drops fall fastest they are the ones that reach the ground first, which is why a shower begins with those loud splats from single drops, each less than an inch wide, striking the ground or the roof or your head. And the shape of a droplet also varies with its size. If it is below .07 inches (2 millimeters) wide the surface tension of the water is enough to hold the droplet in its natural spherical shape. But larger drops are deformed by the air pressure on them as they fall. The pressure below is greatest so their undersides flatten, while the pressure of the air rushing past the insides is much less, so the sides push out, exaggerating the flattening of the drop. What strikes the ground down below is in fact a sort of flattish pancake or hamburger-bun-shaped globule of water, which immediately spreads out laterally—and it is this sideways force that scours the ground.

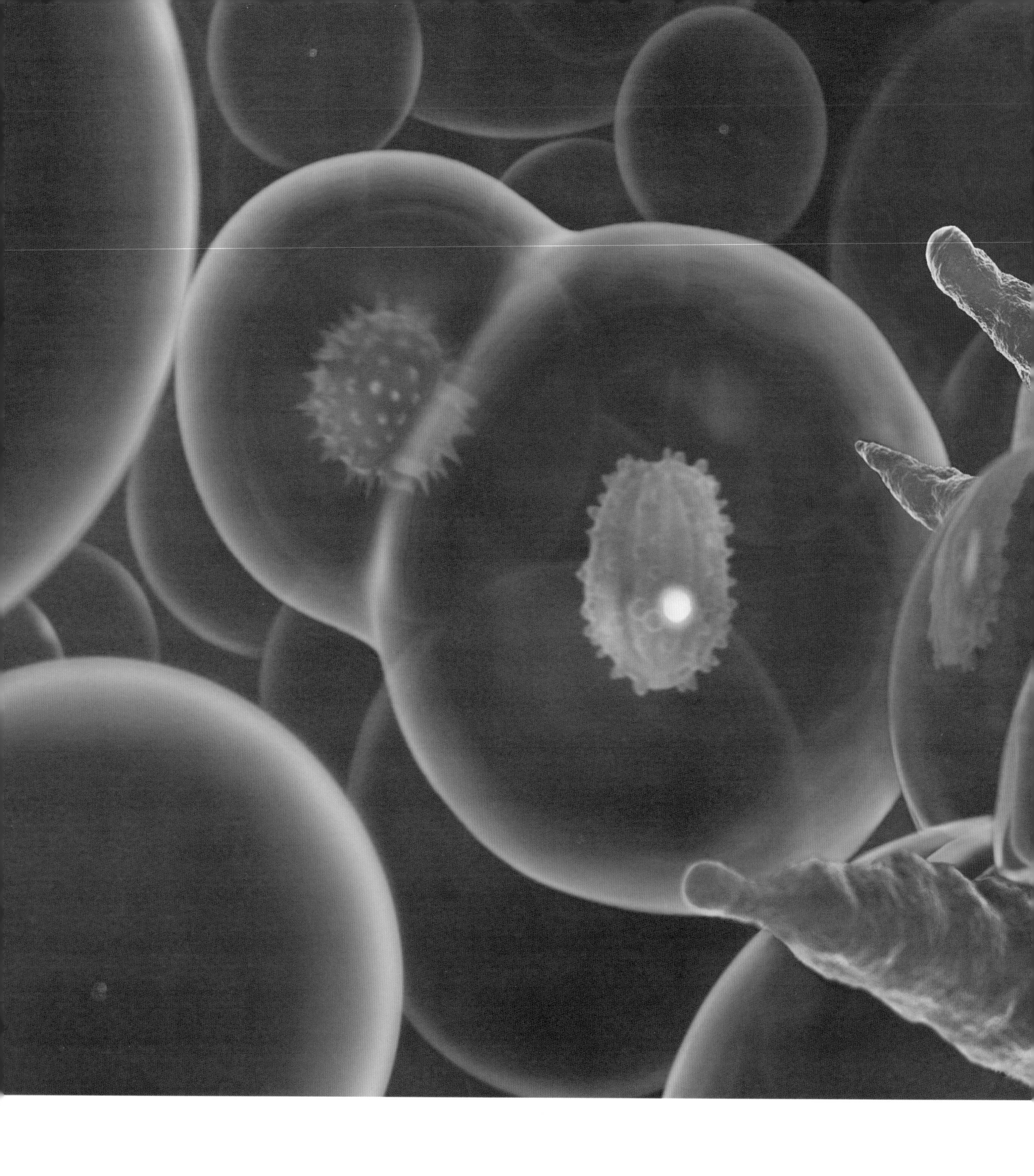

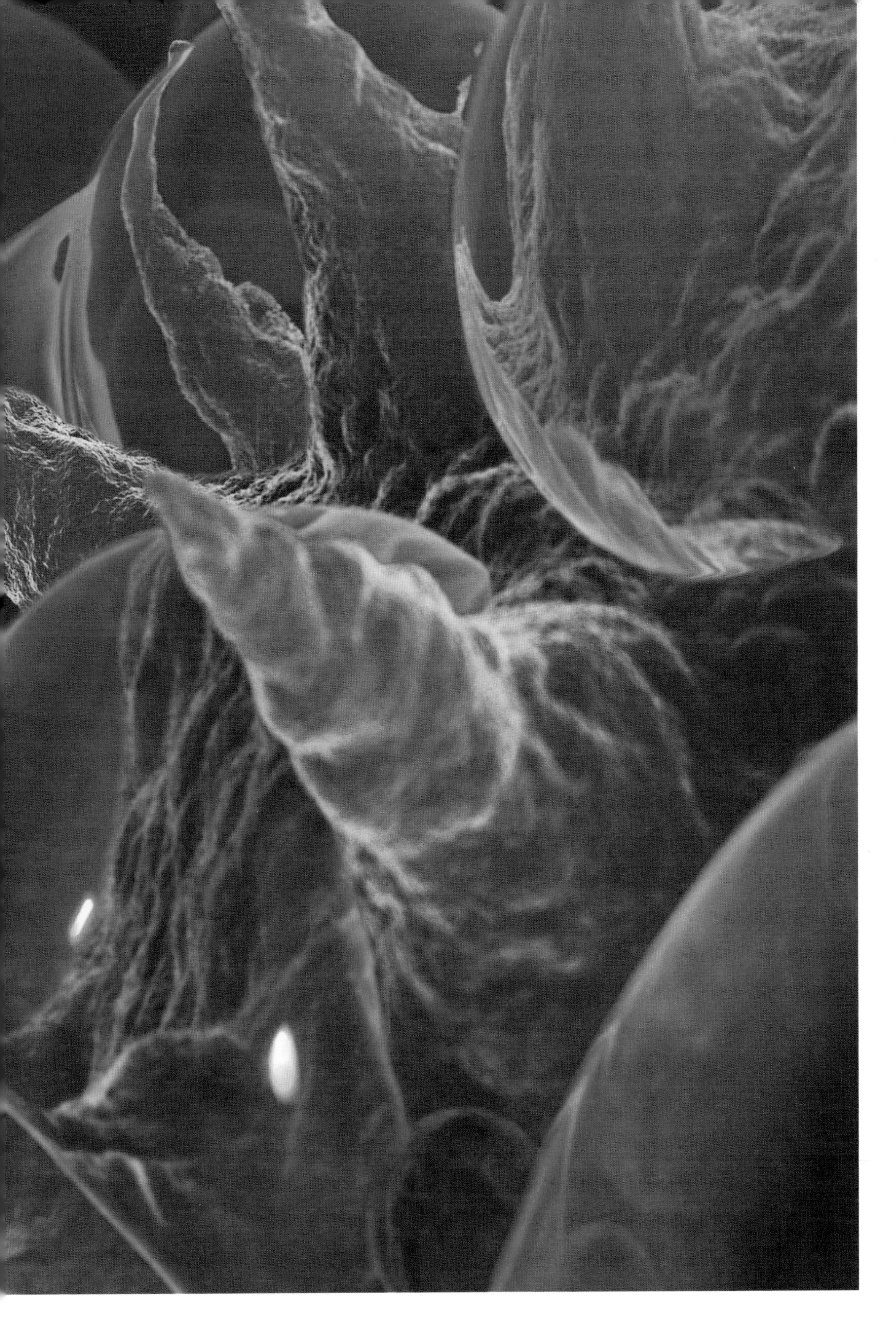

Left: Computer-generated image of raindrops "seeding" as water molecules adhere to microscopic particles in the air.

THUNDERSTORM

BIRTH

It is midsummer and the sun is warming a small area of land. The warmth of the land heats the air above it. The air is still moist from rain the previous night. The conditions are set for the birth of a thunderstorm, a process which begins when a parcel of the warm, moist air begins to rise. There may be another trigger to start the parcel moving upwards faster than the air around it—perhaps a patch of ground is warming more rapidly than the area next to it, or wind is forcing it to move up a mountainside, or some warm, moist air meets colder air and rises above it—but, as we saw in Chapter One, if air is warm and moist it will rise, and that is all it takes to give birth to a thunderstorm.

Above: The warm expanse of the North American prairies provides the enormous heat required to

AWAKENING

As the parcel of air rises up and up, there is less air weighing down on it so it expands and cools. As it cools the water vapor in the air begins the process of condensation, turning into droplets, and a cloud starts to form. As more droplets are made, so the cloud builds upwards—the very process of condensation releases large quantities of heat, which keeps the air inside the cloud warmer than the air surrounding it and encourages it to rise even more. Far below, more air is now needed to replace the air that has risen off the ground and so more moist air is sucked in from around the cloud and the cloud builds up even more. As long as warm, moist air is being sucked up, the cloud will grow until it creates the fabulous, towering columns of a nascent thundercloud. When it hits the top of the troposphere, higher level winds may spread the storm out, to form the famous "anvil" shape of a thunderhead.

The power of the rising air, or updraft, keeps the water droplets suspended in the air, preventing them from falling as rain, as the height of the thunderstorm increases to 39,000 feet (12,000 meters) or more. At that height the air temperature can be as low as –40°F (–40°C), and so ice crystals begin to form. Further down, the water droplets are now growing larger and larger, and heavier and heavier, until they reach the point where the updraft cannot hold them aloft any longer and they begin to fall as rain.

MATURITY

Then comes the very special process that makes a thundercloud different from, and more powerful than any normal cloud. As the heavy rain begins to fall drier air around the cloud is sucked in, causing some of the water to evaporate. This process uses heat and chills the air, making it heavy, so the air around the edges of the cloud begins to fall as a downdraft. The storm now shifts into its "mature phase." With an updraft and a downdraft at its heart, the storm is now in a cycle and becomes known as a cell, the biggest and most frightening of which is known as a supercell.

build huge thunderstorms.

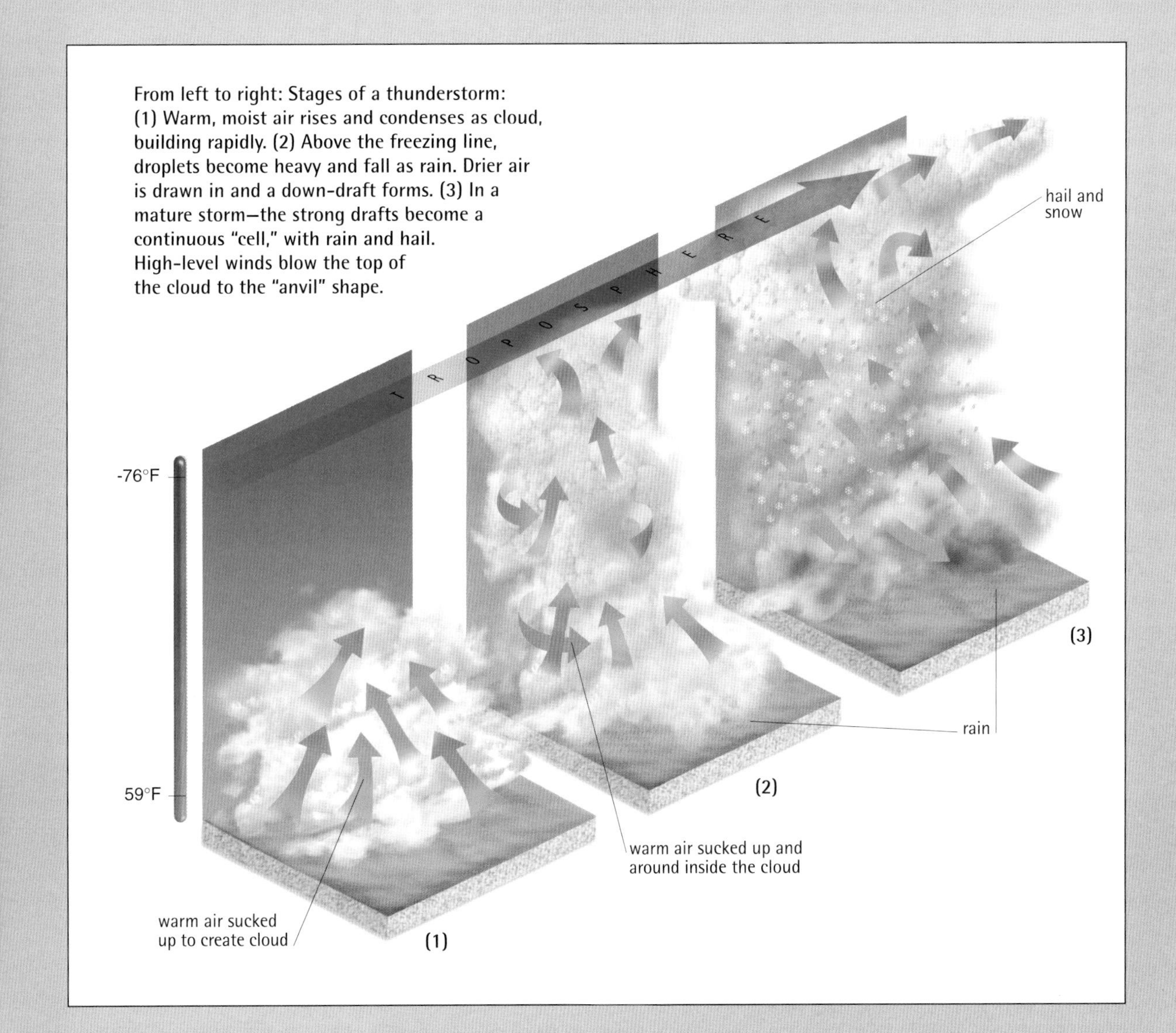

POWER

The updraft below a thundercloud is a phenomenon well known to glider pilots, the bravest of whom sometimes use one to gain height rapidly. But this easy ride upwards is something to be treated with great care, as a tragic story from 1938 reveals. The Rhön mountains in central Germany were a favorite region for gliding contests as they provided strong thermals that could keep a glider airborne all day. One thundery summer afternoon thermals had enabled record-breaking heights of 26,000 feet (8,000 meters) to be reached by some contestants, until five pilots flew their craft into a supercell. All five were immediately drawn violently upwards into the dense cloud, as the high lift ratio of their racing gliders' wings were caught by the updraft, moving up at speeds of 90 miles (150 kilometers) an hour, until they realized their danger, ejected and pulled their parachute cords.

But that was their undoing. Instead of falling gently to earth, the parachutes were lifted by the rising wind and the men were carried on upwards to ever colder regions of the cloud, blown hither and thither, desperately trying to steer with their arms and legs, falling with a downdraft only to be lifted again as they crossed a wind-shear, soaked instantly by giant drops of rain, lashed by huge hailstones. Only one of the five lived to tell the tale. The ultimate fate of the others will never be known. They rose to the freezing top of the cloud 9 miles (15 kilometers) or more above the ground, to become encased in ice where, anyway, the air was too thin to keep them alive.

monsoon

Although the current world number one for annual rainfall is now Llorho, in Colombia, which lays claim to an average of 524 inches (1,330 centimeters) of rain per year, the place that has become synonymous with wetness is the Indian subcontinent during the monsoon. The monsoon's arrival each year can be predicted to the week, and in the build-up to its coming life revolves around the rain. From June to September India is drenched in rain. From October to May it is not. It is simple and straightforward: two huge seasons that govern life, wet and dry. In his affectionate book *Chasing the Monsoon*, Alexander Frater describes the deluge that engulfs India as two wet arms that embrace the subcontinent each year, one flowing up from the Bay of Bengal, across Bangladesh and then turning westwards, and the other driven by a wind coming off the Arabian Sea, from the south-west. These two arms reach up across India, traveling for a month until they meet around the start of July in the north-west of the country, before flowing on into Pakistan.

As the annual date of onset approaches, the meteorological center at Trivandrum on the southern point of India becomes a focus for media attention as it refines and refines its set of predictions for the rains, before "declaring" the onset of the year's monsoon. Prior to the days of radar and satellite, the meteorologists of the British Empire would look out across the Arabian Sea from the vantage point of their weather station in the hills and literally watch for the approaching storm. Today a party atmosphere builds as people line the beach, facing the winds that whip up to gale force running ahead of the monsoon, clutching their umbrellas and heaving a collective sigh of relief as the dark clouds finally arrive on shore and the rain begins.

Four years after Edmond Halley predicted the date of return (correctly as it has since turned out) of the comet that bears his name, he undertook a remarkable voyage through the tropical oceans, which resulted in the first real weather map as we would know it today and his explanation of the flow of the trade winds (see Chapter Two). But Halley also thought carefully about the monsoon. He charted the annual winds, using his very own distinctive, comet-tail-shaped arrows on his map, and he also worked out the basic mechanism that drove it.

In essence it is the same as the model for sea breezes that children learn early on in their geography classes at school. Take a seashore in summer. During the day the land is warmed quickly by the heat of the sun, and the air above it rises through convection and expands outwards. But out at sea the picture is different. Water has a very special property: more heat is needed to raise the temperature of this simple substance than for any other liquid, so the sea heats up more slowly than the land. The result is that the air above the sea stays cooler and, instead of rising, it flows inshore to fill the space left by the warm air that has risen above the land. What you get, then, is a moist sea breeze that gradually builds up during the day and blows inland by the mid-afternoon. At night, however, the situation is reversed. The land cools rapidly, but the bizarre property of water molecules means that it retains its heat much longer. So by night the air rises above the sea and falls over the land, and a dry shore breeze blows out to sea. That simple process of reversal is the mechanism behind the monsoon, but on a planetary scale. For a seashore substitute a continent and an ocean, and for a summer's day substitute an entire year.

What Edmond Halley theorized was that in summer the heating of the Asian continent by the sun created a huge temperature gradient between the land and the ocean. Thus the air would rise over the land and moist air from the sea would rush in to replace it, bringing rain as it

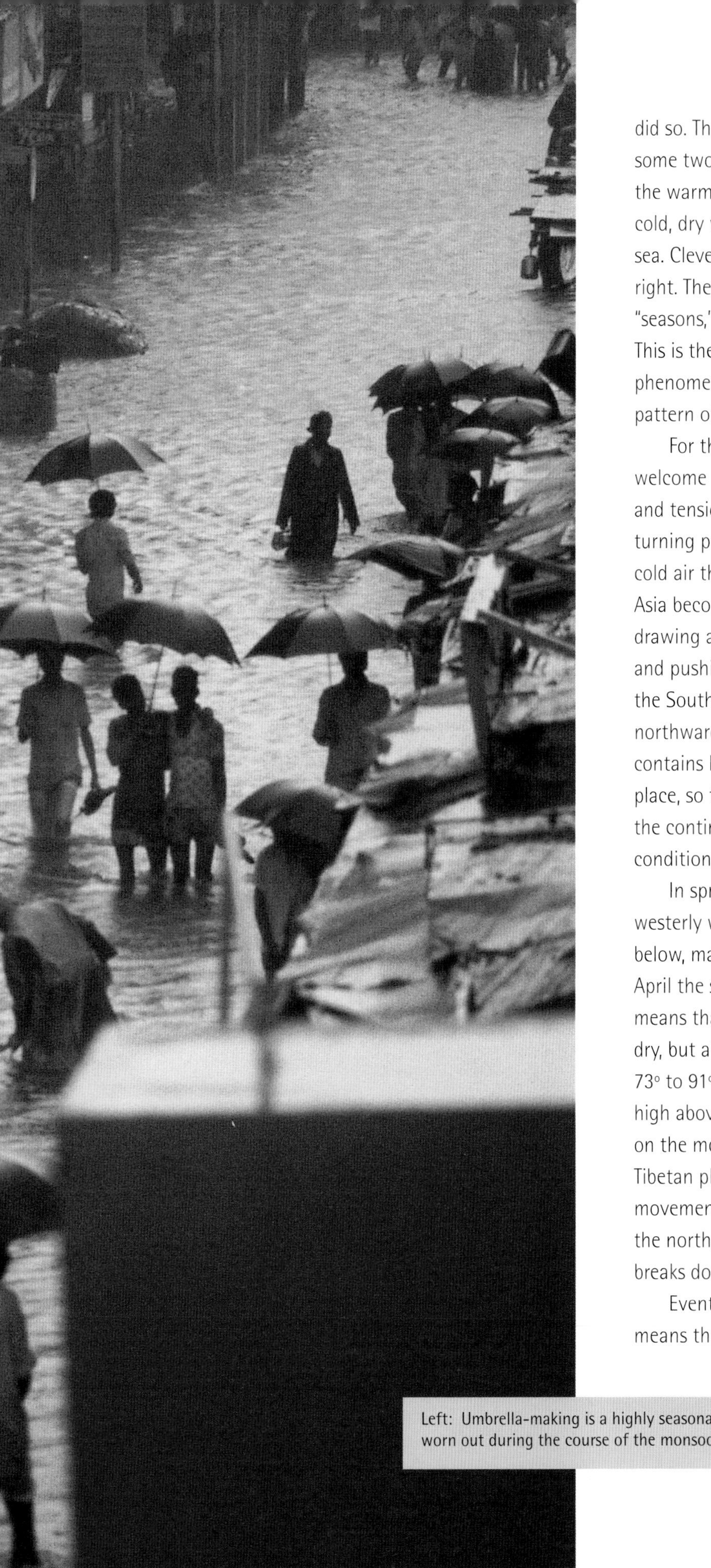

did so. Then as the oceans themselves warmed up some two months after the summer solstice, and the warmth on the land began to fade, so the cold, dry winds of winter would flow back out to sea. Clever chap, Halley. His explanation was right. The result is wet and dry, two simple "seasons," the Arabic word for which is *mausim*. This is the origin of the word monsoon, the phenomenon that dominates the weather pattern of the tropics.

For the people of India the monsoon is a welcome relief from the intense heat, humidity and tension that precedes it, and marks the turning point of the year. Throughout winter the cold air that lies over the continental interior of Asia becomes dense and sinks to the ground, drawing air down from the upper atmosphere and pushing out in all directions, southwards to the South China Sea and the Indian Ocean, and northwards towards Pakistan. Because this air contains little moisture no condensation takes place, so few clouds form. The air traveling out of the continent therefore brings cold but sunny conditions.

In spring things begin to change. High-level westerly winds help to keep the air contained below, maintaining the high pressure, but by April the sun is seriously warming the air. This means that the weather in northern India is still dry, but also hot. Temperatures in Delhi rise from 73° to 91°F (23° to 33°C) from March to May. But high above, the line of the high-level westerlies is on the move, retreating slowly north towards the Tibetan plateau in response to the sun's movement and the shift of the polar fronts far to the north, so the high pressure over the land breaks down.

Eventually, the ever-increasing temperature means that more and more air rises, expands and

Left: Umbrella-making is a highly seasonal business in India, but most umbrellas are worn out during the course of the monsoon.

flows outwards at mid-altitude, leaving more low-pressure areas behind until they come to dominate the continental interior. Then, from May to June, the high-level westerlies jump from the south to the north side of the Tibetan plateau, finally releasing the warm air so that it lifts dramatically from the continent. The low-level, humid south-westerly winds are suddenly sucked in off the Indian Ocean, bringing the monsoon. As the westerlies retreat during the month of June so the monsoon flows in, reaching its maximum extent over northern Pakistan in mid-July. By this time the moist monsoon air covers the whole of India and most of southern and south-eastern Asia. As it moves into the low-pressure regions over the land it rises, forming dense thunderclouds and torrential rain.

For Cherrapunji, nestled against the mountains to the north, the moist air arrives with an astonishing sense of inevitability. The geological shape of the hills has resulted in a narrow valley up which the full blast of the eastern arm of the monsoon is funneled each year until it piles up against the mountains behind. It can go no further. Instead it rises even further, creating more condensation at higher, colder altitudes, and even more rain. The cliffs behind the town are permanently shrouded in cloud, and waterfalls cascade down their face—one is claimed to be the fourth highest fall in the world. In the markets the people wear umbrellas tucked into their headbands.

Four hundred and thirty inches (11 meters) of rain fall here each year, so if it has recently lost its status as the world's wettest place it is not for want of trying. It is a place where you have to love rain or suffer. Joseph Hooker, the famous naturalist who became director of the Royal Botanical Gardens at Kew in London, wrote his *Himalayan Journal* about his time at Cherrapunji during the monsoon, and recorded over 500

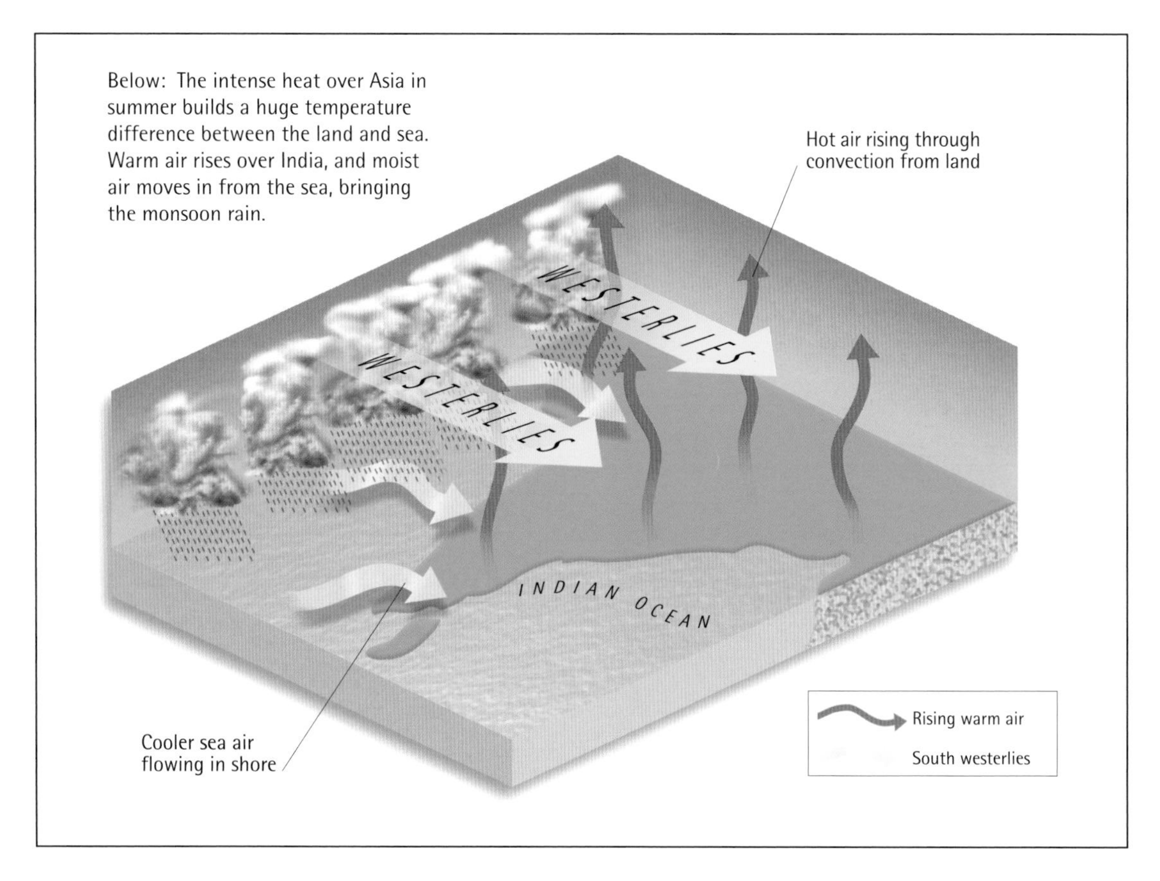

Below: The intense heat over Asia in summer builds a huge temperature difference between the land and sea. Warm air rises over India, and moist air moves in from the sea, bringing the monsoon rain.

Above: Monsoon rain brings a flourishing of growth in an Indian National Park.

inches (1,270 centimeters) of rainfall in one spot during the seven months of his stay in 1850. He found the range of flora growing in this extraordinary climate to be perhaps the widest in Asia, and collected over 2,000 different plants, including 250 different kinds of orchid. For the Victorians stationed at Cherrapunji the climate must have seemed as alien as any that could be dreamt of at home. After the end of the rains, when the water had flooded away down towards what is now Bangladesh, illness and disease brought many casualties in the oppressive and humid weather during the months of September and October. And the weathered graveyard from the days of empire reveals that not a few of the casualties "died of their own hand," perhaps under the intense mental pressure that such extreme wet weather can bring.

the southern oscillation

In 1877 India suffered a terrible famine following a severe drought—because the rains failed. For the monsoon is not quite a perfect clockwork cycle whose regularity can be relied on totally. The rains sometimes come late and they are sometimes weak, and sometimes both. The Government Famine Commission set up in the aftermath of the disaster demanded that the Indian Meteorological Service try, "with the advance of knowledge, to form a forecast in the future," and that it be able to predict the onset and intensity of each monsoon season. At the time it was a task beyond the science that was available, and knowledge had to advance for nearly a century before the system linking the monsoon to the global climate was clearly understood. Early attempts focused on the degree of snow cover over the Himalayas in the winter season that led up to the monsoon. It was hoped this would explain the intensity of the high pressure that forced the dry winds out of Asia, but that was nowhere near the complete answer. More dry years occurred at the end of the century, including another famine in 1899, but five years after that a new director-general was appointed to the service and it was he who made the great leap in understanding how the monsoon fitted into the big picture. His name was Sir Gilbert Walker and he was a classic late-Victorian polymath with an extraordinary range of artistic and scientific interests, such as painting, ice skating, gliding and the study of bird flight. He was fascinated by the performance of the flute and made a small improvement to its design; and he spent ten years working on the properties of the boomerang, countless examples of which he had shipped over from Australia and from which he developed new theories about the nature of gyroscopic motion.

Above: Sir Gilbert Walker gave his name to the pattern of ocean and air circulation that links the weather in the Pacific and Indian oceans.

Walker's contribution to our understanding of the weather was to apply the art of statistics. Long before the days of number-crunching computers, what he had available was a large number of assistants in the Indian Meteorological Service. He set them the task of checking through weather and ocean data from around the world, and of statistically analyzing it and identifying any significant correlation between meteorological and oceanographic events. From this study there emerged a link between the monsoon's severity and time of onset and the relative air pressures over the Indian and Pacific oceans. He found that high pressure over the Pacific tended to mean low in the Indian Ocean all the way from Africa to Australia—and vice versa. He immediately saw this as proof of the monsoon being linked to a global system, and he named this oscillation of

pressure between the oceans the "southern oscillation." Walker spent years building up data to establish that his oscillation correlated with changes in rainfall and wind in the Pacific and Indian oceans, and with changes in temperature in Africa, southern Canada and the USA. Sadly his attempts to use it to predict the monsoon failed and his evidence of a link was called into question by other meteorologists. The problem was that he simply did not have enough data to make the pattern clear, although years later it is now obvious that he was very much on the right track, and today the circulatory pattern that he identified has become known as the Walker circulation. Its name is a tribute to his remarkable achievements, for what he did was far more than provide the first clues to predicting the monsoon; rather, he set out the very foundations for our understanding of the global climate as a whole.

Above: Famine lines in late 19th century India. The failure of the monsoon prompted a drive to understand and predict the seasonal rains.

the christ child

The southern oscillation is linked to the monsoon, but in the years since Sir Gilbert Walker gave it its name it has also become for ever linked to a deeper and more infamous weather event: El Niño. In the last two decades, the El Niño Southern Oscillation (ENSO) has become one of the most studied phenomena on the planet, because it is also one of the most feared.

Shortly after Christmas each year a warm ocean current flows south along the coasts of Ecuador and Peru. Not every year, but occasionally, that current is stronger, flows further south and is very warm. The result is extra heavy rain, which the coastal inhabitants of those countries have always welcomed for the abundance of crops that it brings. And hence the name El Niño (the Christ Child), for the gifts of plenty that it bestows so soon after the Nativity. But El Niño is no longer an event that is welcomed in the rest of the world, because today it is associated elsewhere with severe flooding, severe drought, reversal of normal rainfall patterns and human misery. In 1998 one of the strongest El Niños on record brought great destruction to Peru, where 30,000 homes were destroyed and a 93-mile- (150-kilometer-) long lake appeared almost overnight across an area of desert that had been dry for over fifteen years.

To understand a major El Niño event, as it is known, we need first to understand "normal conditions." Normally there is a large area of high

Right: The very strong El Niño event of 1998 brought catastrophic flood damage to Peru.

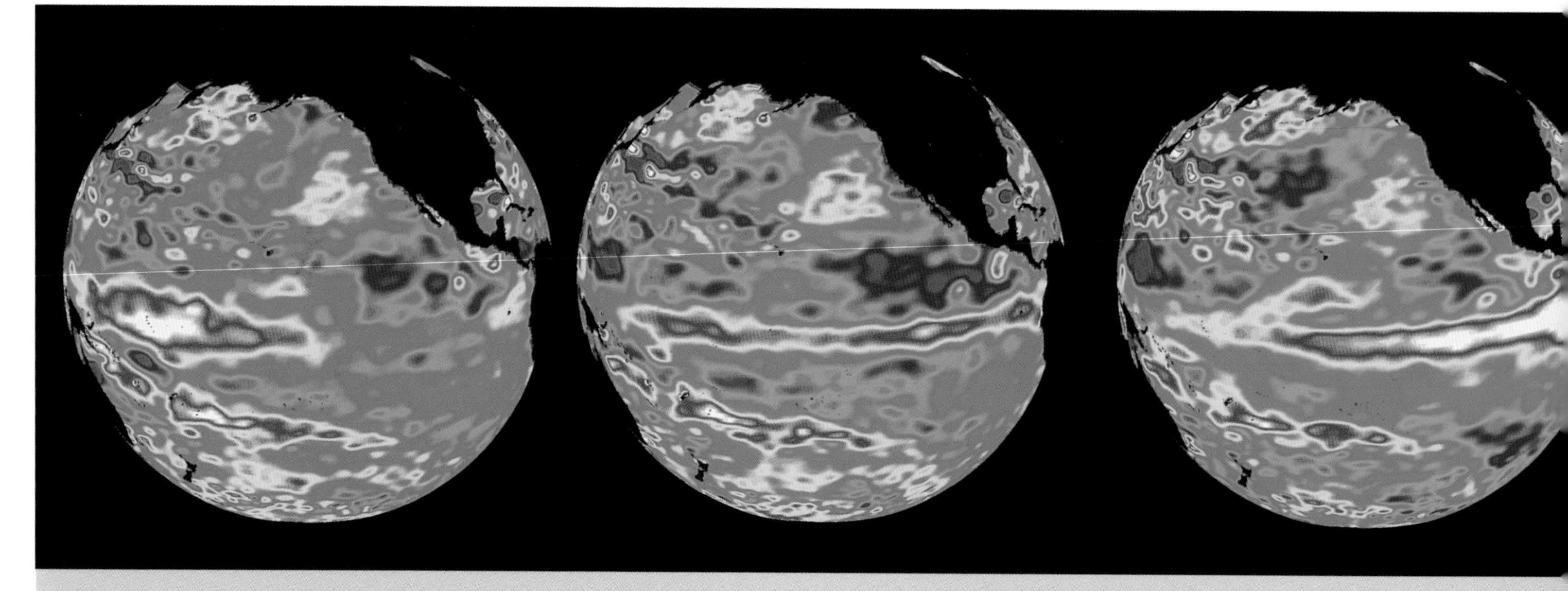

Right: This sequence of satellite images show the 1997 El Niño developing as it moved eastwards across the Pacific (colored in red). The images from left to right were

pressure sitting over the eastern edge of the Pacific Ocean, just off the South American coast. From this high-pressure zone the southern trade winds blow towards a large area of low pressure that is settled over Indonesia, on the other side of the ocean. These steady winds are strong and they drag the cool water that lies off South America westwards with them, so that it warms by contact with the atmosphere and by the heat of the sun. And over the vast distance that is the width of the world's largest ocean this westward flow of water actually raises the sea level in the western Pacific—it is some 16 inches (40 centimeters) higher than it is next to the South American coast. The result is a thick layer of warm water over the western Pacific, and a mild deeper current that flows towards the east, known as the countercurrent, which gently brings the warm water back, to cool, rise up off South America and start again. In the sky above, meanwhile, this body of warm water evaporates and moist air rises, adding to rains such as the monsoon. Further aloft, and now drier, the air is carried by the fast-moving upper-level winds to the east, where it cools and descends, adding to the high-pressure zone off South America where the cycle began. This, in essence, is the Walker circulation, and this is "normality."

An El Niño event is different, however. All it takes to trigger it is a slight relaxation of the trade winds, when a feedback loop immediately sets in. The weaker winds mean that the mass of warm water in the west suddenly surges eastwards, strengthening the countercurrent and forming a warm layer of water much closer to South America than usual. That warmth in turn warms the air above, which lowers the pressure of the atmosphere, and so the winds weaken even more. So more warm water surges eastwards and suddenly there is a reversal. The whole of the Pacific is effectively covered with an extra layer of warmth and the winds change direction, settling into a new long-term pattern in the opposite direction. Far to the west it means that the easterly winds that drive the moist air off the Indian Ocean, carrying it up to curve across the Equator and build to the monsoon, are also weakened, and the monsoon with them. Far to the east, down the coast of South America, the increased thickness of warm water means more

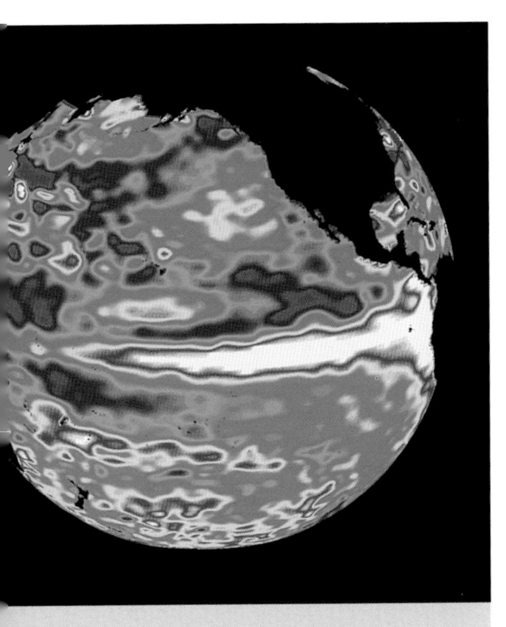

taken in April, May, June and September.

evaporation to the atmosphere and so more rain above. For the farmers inland the rains can be a boon, but for the fishermen who haul in the world's prime anchovy catch each year they are a disaster. Normally the cooler water rising off their coast, waiting to be blown west on the trades, brings rich nutrients from the deeper ocean and supports some of the world's most abundant fishery areas. But with El Niño the nutrient-rich, cooler waters the fish need are held too deep by the layer of warm water above, so the surface sea becomes hostile and the fish flee south or die. In today's modern interlinked economies the failure of the anchovy catch means a massive depletion of the world's supply of fishmeal for animal feed.

The effects of a major El Niño can be widespread. Such a large shift of temperature, pressure and wind patterns is felt in many ways. One of the main factors is the change in the path of the subtropical jet stream, the high-level wind that flows like a belt around the northern hemisphere at about 30 degrees from the Equator, and whose great speed pushes and drags smaller weather systems along beneath it. The large area of low pressure that develops over the western Pacific draws the jet stream south, so that it carries the thickening Pacific clouds eastwards to pile up more storms against the coast of California and Mexico. But at the same time the repositioned jet stream brings benefits, because it cuts off the tops of

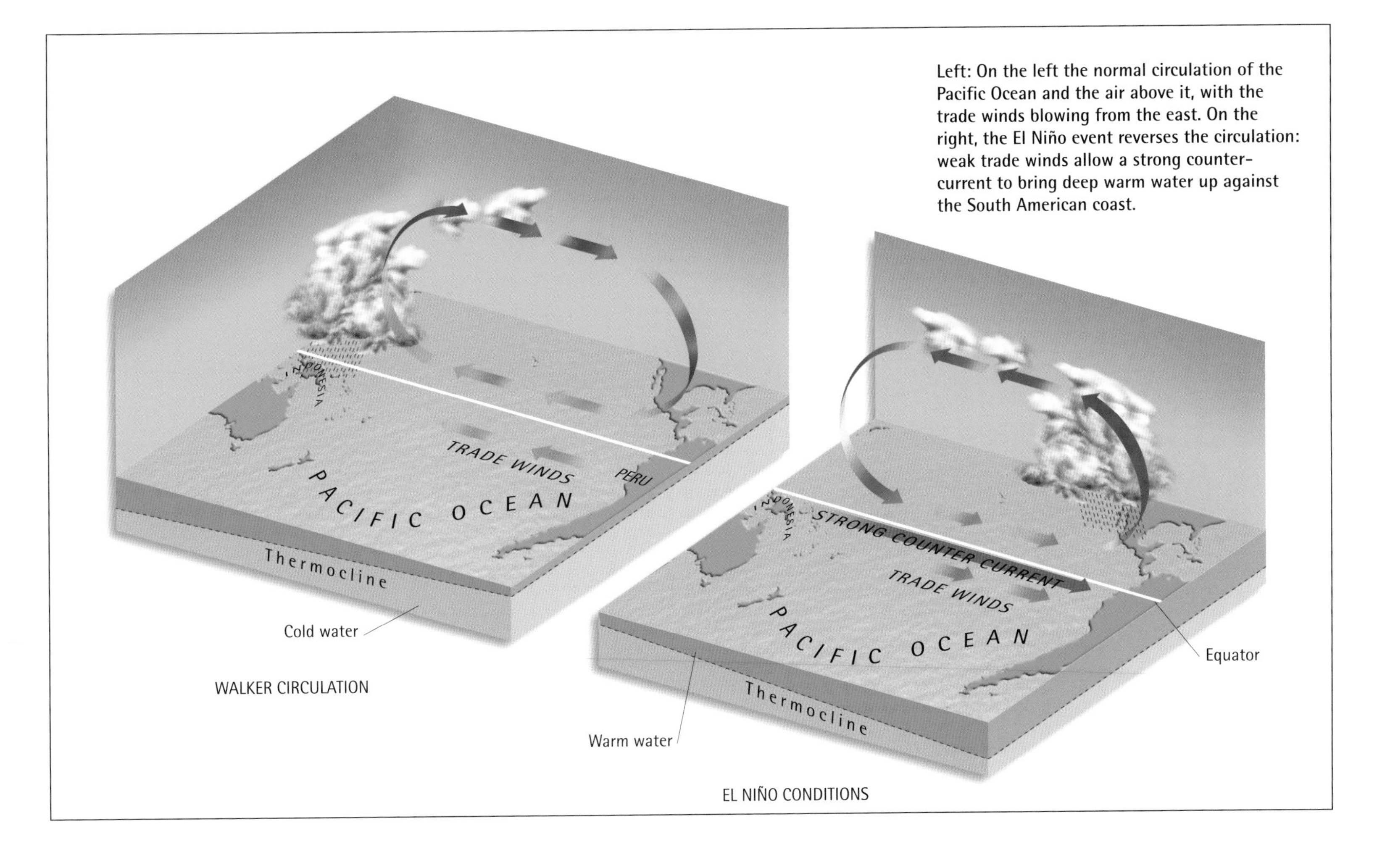

Left: On the left the normal circulation of the Pacific Ocean and the air above it, with the trade winds blowing from the east. On the right, the El Niño event reverses the circulation: weak trade winds allow a strong counter-current to bring deep warm water up against the South American coast.

developing Atlantic storms which tends to prevent them building up to become hurricanes. Meanwhile, in the recent El Niño event of 1997 and 1998, which is the best-documented and studied example of the phenomenon since scientists became seriously aware of its implications, East Africa experienced dramatically higher rainfall, boosting its coffee harvest, while Brazil suffered drought and losses to the same crop. Central Asia and the north-west USA and Canada experienced heatwaves, while some countries in central Europe suffered flooding. The degree to which all of these factors are cause-and-effect will be debated for years, but scientists can at least now clearly see that there are links between these huge shifts of global weather.

The reversal of the Pacific flow which lies at the heart of El Niño can last for a couple of years before conditions build in the reverse direction, and then La Niña (the little girl) appears. Out of twenty-three El Niños that are now thought to have occurred in the last century, fifteen of them were followed by La Niña. When this happens, instead of a comfortable return to the normality of the Walker circulation in the Pacific, the ocean and sky overcompensate, the low pressure deepens over the western Pacific, the trade winds become exceptionally strong, a huge upwelling of cool water emerges from the deep off the South American coast and warm water piles up in the west. The Indian monsoons become stronger, the jet stream shifts away and weakens, far to the north cooler air swings down from the Arctic into North America and a different set of extremes afflicts the globe. This huge ebb and flow across the Pacific, of volumes of warm and cool water that are almost unimaginably vast—you could think of it as a wave of warm water the size of Australia—is like a giant see-saw sloshing slowly from one side of the planet to the other, perfectly fitting its name, the southern oscillation.

Above: The effects of El Niño are widespread. California experienced major storms.

the great ocean journey

In India, around the time of the European autumn, the sun moves lower in the sky once again and the continental air cools, weakening the low-pressure systems that have developed. The monsoon withdraws southwards and, as the surface of the earth cools down, the high pressures and dry winds return to start the whole annual cycle once again. The water has now poured off the Indian subcontinent, bringing the annual flooding across Bangladesh, which can cause such extensive misery for its impoverished population. And then it flows out into the Bay of Bengal to join one of the longest and most exciting journeys possible, traveling even through time to the other side of the globe.

For the water is re-entering the ocean and setting forth into a newly discovered and mysterious world of motion, part of a one-thousand-year voyage that will end with it raining down again in Europe.

Imagine a time, long ago, when a sailing ship was nearing its home in western Europe, blown along on the Atlantic westerlies that could sometimes violently, sometimes gently, carry sailors back to their loved ones after journeys of courage, adventure and privation that few of us can comprehend today. Imagine, perhaps, a sailor on one of Sir Francis Drake's privateers, possibly the great sea dog himself, singing quietly as he sluices his face to freshen up in the early morning, musing on the clean clothes his newly plundered wealth might afford him when he reaches Plymouth in a few days' time. Imagine the dirty water tipped unceremoniously over the side of the ship to mix and swirl away in the foaming brine of the ocean. Well, that very water may rain upon you tomorrow. For Sir Francis Drake's ablutions would have certainly become a tiny part of the most formidable water cycle on

Earth—the great thermohaline conveyor.

It can begin in many places, for it is a continuous loop, but let us start in the North Atlantic Ocean. The Gulf Stream, the ocean current that flows from the Gulf of Mexico up the eastern coast of the USA and across the Atlantic Ocean to western Europe (and of which more later in this chapter) carries warm, salty water from the gulf up to the north-eastern Atlantic. Indeed the North Atlantic is the warmest and saltiest of the Earth's oceans. But up near the Arctic the water cools, becoming heavier, and at the same time sea ice forms above its surface. As the ice forms it draws on the pure water to build its frozen structure, while any salt is left behind in the ocean lapping underneath. Now the water is even saltier, even colder, and so even heavier. It begins to sink, drawing more warm water from the south-west to replace it at the surface. This mass of cold, dense, salty water plummets to an extraordinary depth in the ocean, joining what oceanographers catchily name the North Atlantic Deep Water, which fills most of the deep Atlantic basin, to begin a long, sluggish journey south through the cold darkness of the world's ocean depths until it reaches the other side of the globe.

The long journey is not straightforward: it is like a set of pools in a fountain, where the water enters, travels round and round for some time and then flows down to the next level. So the water in the deep-ocean circulation moves round and round in a series of what are called "gyres"—rotating pools perhaps the width of a sea—with water quietly remaining in one such pool for years before slipping out and moving on to the next one. The path the deep water takes, driven by ever-so-slight changes in temperature and salinity, sweeps back across the Atlantic at a depth of 3,000 feet (1,000 meters) and below,

Below: The basic flow pattern of the thermohaline conveyor, which circulates all the water of the world's oceans over a period of a thousand or more years. Blue shows the cold, deep-water flow, while red shows the warmer surface current.

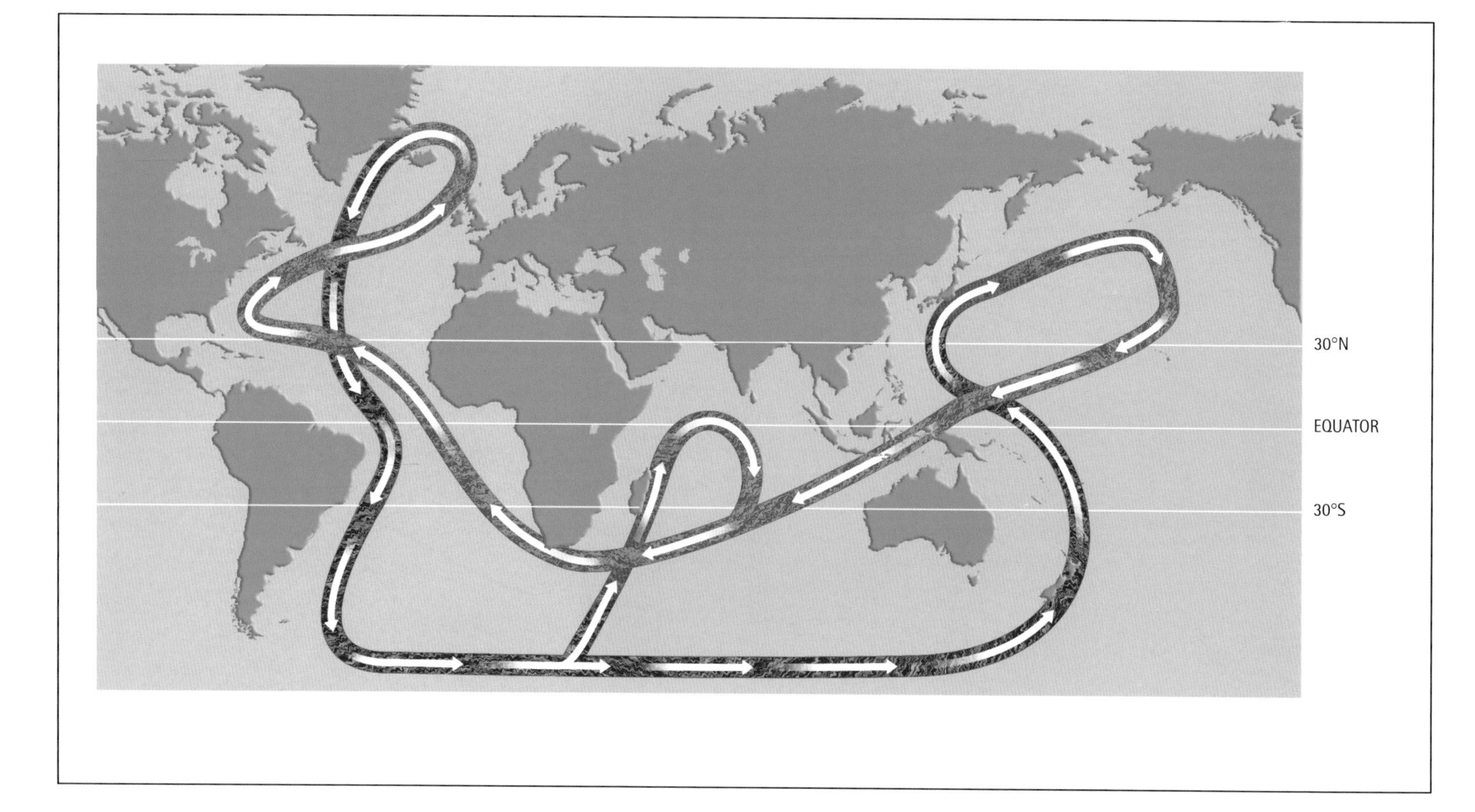

slowly heaving across towards South America, with our tiny droplet of water from Sir Francis Drake's ship. Having circled and circled for perhaps a century, it possibly reaches the seas off northern Brazil at some time in the decades after the Great Fire of London. Then it heads south where, perhaps as the French Revolution was at its height of terror, our molecules meet up with the cold waters descending off the Antarctic coast which creates the Antarctic Bottom Water—the other great source of sinking water that drives the conveyor. From there the conveyor splits in two, with one branch passing on to circle Antarctica and then travel the ocean south of Australia, where it will continue on for another few hundred years before flowing up to the northern Pacific Ocean and then through Indonesia. On the other branch the water journeys north and east, sliding up the coast of Africa as the era of the Victorian explorers gets under way, gradually warming as it does so, decade by decade until, perhaps around the time of the end of the Raj and the emergence of the independent state of India, it rises to the surface of the Indian Ocean in time to become part of the vast evaporation of moist air that creates the rains of the monsoon. And at the end of that summer it could flow back down the rivers of what is now Bangladesh, back into the ocean, to begin a much swifter return journey at the surface, this time of only fifty years.

The surface currents from the Indian Ocean travel back south where they are joined by the surface currents flowing in from the Pacific. (Those waters would have been part of the longer branch of the conveyor, waters which began their journey hundreds of years before, but which surfaced in the northern Pacific and then flowed back through Indonesia.) Together these giant streams of water take a few decades to travel round the Cape of Good Hope at the southern tip of Africa, lingering for years at a time in spiraling eddies and gyres, before ending up in the Caribbean, eventually to be picked up again by the Gulf Stream and swept at an extraordinary speed back up towards the northern Atlantic. It is an epic journey, but it is not yet over. For the power of the waters in the Gulf of Mexico is worth a brief examination on its own.

flood in the gulf

Recent research at the National Oceanic and Atmospheric Administration (NOAA) in the USA suggests that the ocean circulation itself can have a very direct effect on local weather. In the warm waters of the Caribbean there is a surface current that circles the Gulf of Mexico over a period of about a year. Known as the Loop Current, it is a precursor to the Gulf Stream itself. As it slowly circles the region, smaller eddies of warm water "spin" off from it, turning relentlessly and trapping water which heats with the sun creating deep pockets of warm water that spiral gently in one spot. Above these patches of extra-warm water the rate of evaporation becomes intense and moist air hangs above them for long periods of time. That in itself would be no great worry but for the Caribbean phenomenon of the hurricane. It is now thought that some of the worst hurricanes of recent years were intensified when they happened to pass over the heat energy of one of these warm-water reservoirs and were fueled to build to greater strength by the warm, moist air and intense evaporation beneath them. The hurricane winds take heat from the water but, because the reservoir of warm water is so deep, it keeps on supplying more energy in the form of evaporating water. This likely link between the ocean currents and the most feared of storms has sparked a rush of new investigations, in the hope of yet more accurate predictions of disaster.

Right: Onlookers watch the waters of the Choluteca River overflow, causing huge death and destruction. This flash flood occurred in Honduras on October 31, 1998 in the aftermath of Hurricane Mitch.

water not wind

At the end of October 1998 Hurricane Mitch became the most deadly hurricane to have struck the American continent in over two hundred years. The "Great Hurricane" of 1780 had killed more than 22,000 people in the Caribbean when it struck at Barbados and Martinique, but that had felt like history. The final death toll of Mitch will never be exactly known, but it stands today at well over 11,000. Its primary impact was on the fledgling nation of Honduras, and on Nicaragua next door—more than three million people were made homeless, and the damage estimates are at $5 billion, and still rising. But most of the death and destruction was not brought about by the 200-miles- (300-kilometer) an hour winds. What is often not understood is that when hurricanes strike their most deadly aspect is the water they disgorge. On the coast, the winds had whipped up

Left: Devastation wrought by Hurricane Mitch in Santa Rosa, Honduras, in October 1998.
Above: Satellite image of Hurricane Mitch as it settled with full force over Honduras and Nicaragua.

30-foot (10-meter) waves that removed fishing villages in their entirety and sent boats to the bottom of the ocean. But even when the gales had lessened the rain continued to fall. The rainfall from Mitch was extraordinary, with 20 inches (half a meter)—the equivalent of five months' normal rain—falling in a single day around Tegucigalpa, the capital of Honduras. The consequence was flash floods from the rivers flowing down from the mountains behind it, and mudslides that resulted in a torrent that poured right through the center of the city, overtopping the bridges that straddled the river and carrying debris, rubble, cars, buses, dead animals and human bodies. Throughout the northern part of Honduras, bridges were washed away, and flood waters ruined entire crops leading to starvation the following year. In the south, the Rio Choluteca cut a completely new valley down the mountains to the Pacific Ocean, taking villages and people with it as it carved its way. Water was the real killer.

the gulf stream

After lingering in the Caribbean the water we have been tracking shifts out of the Loop Current to slip across into the area that is the source of the Gulf Stream, probably the most famous ocean current in the world. This current of fast-flowing ocean sweeps up the eastern coast of the USA from the Florida Keys. It was first remarked upon in historical records in 1513 by the Spanish explorer Juan Ponce de León, who had been seeking the "fountain of youth" but settled for Florida. He had been sailing south through the Florida straits when, despite a full breeze behind him, he found that his vessel was moving backwards in the ocean. He realized that he had stumbled upon a very strong current, put it down to the typical behavior of the seas as they narrowed through the straits he was traversing and steered his ship close to the shore to get out of its path. What he did not realize was that this flow of water, measured today at 1050 million cubic feet (30 million cubic meters) per second, was setting out on a journey that would have carried him back to Europe at speeds of up to 100 miles (160 kilometers) a day. The pilot who had sailed with him, Anton de Alaminos, is credited by some as being the first European deliberately to use the powerful flow when he followed its rapid course six years later on a journey from Mexico back to Spain. This short cut to travel was probably well known to the Seminole Indians who inhabited Florida long before the Europeans arrived, and it rapidly became well known to European sailors. However, many kept its properties a guarded secret as they hoped to use it to their advantage and get wealth from the New World back home more quickly to their king or queen or commercial sponsor. For the local people of Florida a beneficial side effect of that popular strategy was a profitable wrecking trade that was built up to prey on the victims of the fast-moving but often treacherous waters of the Keys.

The secrecy surrounding the Gulf Stream put those who operated ocean-going services, but knew nothing of the current, at a serious disadvantage. The official mail ships that plied between Britain and her American colony were among the losers, taking two weeks longer on their ocean crossings than the merchantmen who passed by them. On the voyage to America those who knew the currents would steer a course far to the south, out of the flow of the Gulf Stream, even though it looked a far longer route on the charts, while the king's mail ships would loyally follow the charts and steer against the current.

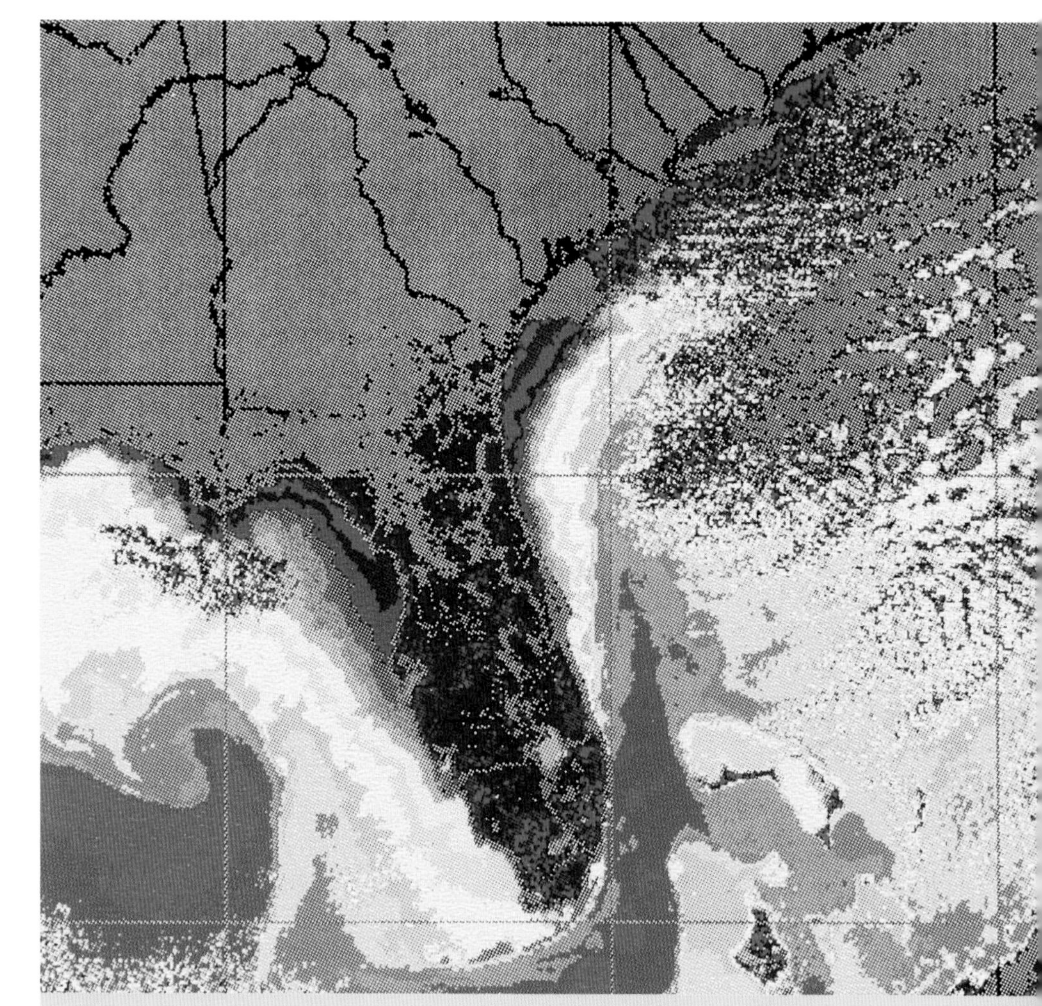

Above: Satellite measurements of the sea surface temperature around the Florida Peninsula, where the Gulf Stream begins its flow.

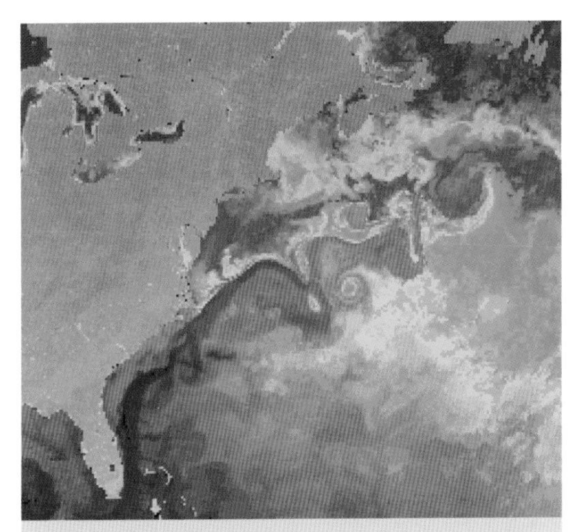

Above: The pattern of eddies and gyres that mark the full complexity of the Gulf Stream as it moves across the North Atlantic.

For twenty years, leading up to the eve of the American Revolution, the deputy postmaster general for the colony was none other than Benjamin Franklin, and he it was who endeavored to find out why the mail packages were so slow. His cousin Timothy Folger was a whaling captain who knew all about the warm current because whales would gather at its edges to feed off the fish and squid that swam in the seas close to its boundaries. Folger told Franklin of incidents when he had advised the mail ships to alter course, only to find his advice rejected by captains who were paid to know better than mere whalers. Folger drew up a chart of the "stream" for Franklin, who had it printed and presented to the postmaster general in Britain—who ignored it. Franklin himself, however, remained determined to identify what was going on in the ocean. On a voyage out from England in 1775 he took temperature measurements from the side of his ship and discovered that the current was much warmer than the ocean on either side of the stream. He also observed that the waters within it were blue, while the ocean on either side was gray, and that whales did not swim within it because it appeared to carry little food. He even found time to continue these scientific observations every day throughout his dramatic voyage to gain support from the French for the revolution the following year. His studies were fruitful, and he pretty much correctly explained the current as being due to the "accumulation of water on the eastern coast of America between the tropics," driven there by the trade winds. He also suggested the name "Gulf Stream."

The Gulf Stream is now one of the most studied ocean currents in the world. A hundred years ago scientists began to realize the importance of temperature and salinity in driving the current, and the action of the Coriolis effect (see Chapter One) in shaping and containing it. Throughout the last century an array of scientific instruments have come into being with which to measure every aspect of the Gulf Stream: the neutrally buoyant float to measure current flow; current meters; SOFAR floats and RAFOS floats to capture acoustic signals; echo sounders; tomographic transceivers; and in the last twenty years, infrared and microwave sensors on satellites that are permanently stationed in orbit have been able to measure the current's depth and motion to within inches. Today the picture of the Gulf Stream is precise and wondrous in its complexity. It is a vast ocean river, constantly turned from its path by the Coriolis effect of the spinning Earth, but counterbalanced by a difference in sea-level height of 3 feet (1 meter) from one side to the other, which creates a huge mass of water heaving the other way. It is not one current but a whole circulatory system within itself, with cold and warm rings and eddies that form, spinning away from it, out and back. At the surface of the ocean it has a secondary, gentle current which returns its flow via a pathway further south, from the North Atlantic to the Caribbean. At the same time it is an essential part of the global circulation of the deep-water ocean. And its influence on our climate and weather is legendary.

warm weather

It is a common misconception that the warm waters of the Gulf Stream sweep up across the Atlantic to reach the western shores of Europe and the British Isles, and that this warmth provides the region with generally milder weather than they would otherwise experience at a latitude so far north. In fact, the current has no direct effect upon British weather at all. What happens is that the intense warmth of the Gulf Stream releases unimaginable quantities of heat and evaporating moisture from the surface of the ocean in a tightly confined area along the length of the North American continental coastline. The current is approximately 50 miles (80 kilometers) wide at its maximum, but can experience a temperature difference of up to 18°F (10°C) with the ocean on either side. At its peak flow south of Nova Scotia it is shifting 5300 cubic feet (150 million cubic meters) a second, or 540 billion tonnes of water per hour, whichever statistic is easier to comprehend in its vastness. The latent heat locked up in such volumes of water has been calculated to equal a hundred times the entire energy consumption of the human race: a humbling thought. That is the kind of energy that can change the weather of northern Europe, for the warmer air that is created in the atmosphere then travels the Atlantic to govern the formation of the cyclones and storms and weather fronts that roll in to greet us.

The warm Sargasso Sea is a volume of water trapped inside a huge circulating eddy that spins out from the Gulf Stream. Warm air builds up above it, and the Gulf Stream divides this warm air from the colder air that forms along the north-east continental shelf of the USA. It creates the oceanic equivalent of a weather front, and at certain times of the year this boundary is the site of some extraordinary weather phenomena. In winter large volumes of cold Arctic air slip gently down through Canada and across the eastern USA, before sliding out to the ocean and encountering the warm water of the Gulf Stream. This results in a vigorous flux of moisture and heat, with huge "chimneys" of steam rising like columns from the surface of the sea to reach the bottom of the clouds. The energy bound up in these "steam devils," as they are known, is huge. The region of ocean in which they form has the equivalent of a modern nuclear power station pumping energy into the atmosphere above each square mile of sea. The results of these astonishing updrafts can be as varied as the fueling of ice storms over New England, or the building of Atlantic cyclones that drench Britain.

All along the so-called "north wall" of the Gulf Stream (the north-west sector of the current) warm, moist air rises and flows out across water that is much cooler—cooler than its dew point—so condensation occurs immediately and large banks of fog are created. For half the year fog also forms where the icy Labrador Current swings south to strike the Gulf Stream, and is intensified by air from the south that picks up moisture as it crosses the path of the warm current, only to arrive in icy conditions on the other side. The icebergs that calve from the Greenland glaciers (see Chapter Four) float south, adding to the formation of fog on their way, only to emerge suddenly into the Gulf Stream where they are caught and melt before they can move further south.

the wild atlantic

Whether it remains flowing in the sea itself, or is evaporated high into the atmosphere to become part of the storms that work their way across ocean, the journey of a molecule of water across the Atlantic will be both eventful and dramatic.

The North Atlantic remains one of the wildest regions of meteorological activity on the planet ●

It will also be largely unpredictable. The North Atlantic remains one of the wildest regions of meteorological activity on the planet and, despite the extraordinary advances in measurement and computing since the Second World War, it continues to surprise and defy the skills of forecasters in the world's most advanced economies. The Atlantic is the source of much of Europe's weather, and anticipating its effects has become essential for many modern human endeavors, as well as for the more traditional necessities of agriculture and shipping. But barely a decade goes by without a new influence being identified that had hitherto had an unknown effect. The eddies and rings of the Gulf Stream and the existence of the thermohaline conveyor are just two relatively recent additions to the scientists' list of phenomena that need to be taken into account. In the last fifteen years science has begun to face up to the complexity of the weather systems that it has to deal with, and a remarkable shift in thinking has occurred with the emergence of Chaos Theory.

In 1960 Edward Lorenz, a meteorologist at the Massachusetts Institute of Technology, was working on a project to simulate weather patterns with a computer, with the aim of improving the ability of forecasters. His computers were very simple, with little memory, but they were able to cope with calculating the interaction of significant phenomena such as tornadoes, hurricanes and easterly or westerly winds. From the consistent patterns that emerged from his trials Lorenz developed predictions of what might happen next in the model. But when running a repetition of a particular pattern of events he made a profound discovery. For this second run of the computer he punched in a set of variables (like temperature or wind) which were very slightly different from those in the first run. He expected the model to come out the same but to his astonishment it began to diverge from the first outcome, and after it had been run for a few model "months" the resulting weather pattern was dramatically different—so different that he realized that the tiny variables had had a totally unpredictable effect. What Lorenz had stumbled upon was the fact that very small variations in the outset of a climate system can result in radically different outcomes. This is at the core of Chaos Theory, which essentially says that such complex systems are, by definition, predictably unpredictable. Lorenz concluded by coining the now-famous analogy of the "butterfly effect": he suggested that, taken to its extreme, the influence of a butterfly flapping its wing and creating a tiny eddy of air over Brazil, might eventually be felt in the intensifying of a hurricane over the Atlantic. In reality it would take a pretty big butterfly to have a measurable influence on the formation of a hurricane, but the principle is sound.

Lorenz also concluded, from repeated runs of his model, that there were some basic rules of "chaos" which meant that no conventional forecast could be reliable beyond ten days to two weeks in advance. And further, it meant that forecasts based on similarities with past weather patterns were fundamentally doomed. But science has moved on, and today's meteorologists have embraced chaos and begun to use it to refine their forecasts to an extraordinary degree. In the early 1990s a new technique called "ensemble forecasting" was developed. What happens is that the forecasters set many different medium-range forecast models going, all with very slightly different starting conditions, to allow for a few "butterflies" that have not been observed out in the atmosphere. In a classic example of the use of this technique the European Center for Medium Range Weather Forecasts (ECMWF) ran their computer model fifty times for the conditions building up in the Atlantic and approaching France in late December 1999. The heart of the Atlantic is recognized as being a

Above: Satellite image of the great British storm of October 1987 in which wind speeds of over 100 miles (160 kilometers) an hour were recorded.

rapidly changing cauldron of violent weather, and yet it is also a meteorologist's blind spot, for there are few weather stations or measurements from ships or balloons. The normal forecast based on the known prevailing Atlantic conditions showed no particular threat of low pressure developing over France. However of the fifty model runs, each with the same basic starting conditions but each with slightly different variables added in, five developed vigorous low-pressure systems right over Paris. That was enough for the forecasters to issue a warning of severe weather, and it was as well that they did. Winds of more than 125 miles (200 kilometers) an hour tore across the continent and Paris suffered its worst storm for fifty years. Spires were torn off Notre Dame cathedral, many of the elegant, tall glass windows in the palace of Versailles were smashed and 10,000 trees were uprooted in its gardens; it was even feared that the Eiffel Tower itself would snap. The storms produced one of the worst ever insurance losses across Europe. Had meteorologists been able to apply the same approach in the lead-up to the famous storm across England in October 1987, they might well have been saved the embarrassment they suffered then, when suggestions of hurricane-force winds were dismissed.

Above: Millions of mature trees were lost across southern England during the famous storm of October 1987.

father of meteorology

The development of the fundamental principles that govern the science of meteorology today can be laid at the door of one remarkable man and the small team of dedicated scientists who worked with him over a period of barely five years at the end of the First World War. Vilhelm Bjerknes was a Norwegian physicist whose passionate aim was to reach a total understanding of the physical principles involved in the forces of the weather. As we know today such absolute knowledge is almost certainly unattainable, but Bjerknes, in his day, made breakthroughs that have brought us very close to it. In the years before 1912 he had attempted his study in both Sweden and the USA, but could not lay his hands on the resources necessary for such a Herculean task until he was invited that year to be the director of a German government-funded research institute in Leipzig. But the outbreak of war meant that the demands made on him for regular weather forecasts for the military prevented him making much useful progress. In 1917 he accepted an invitation from Bergen Museum to be a professor of physics at a brand-new geophysical institute which was part of a grand plan to build up a university in that southern Norwegian town. Together with his son Jacob and a small team of colleagues he embarked on a program of research that was to revolutionize meteorology.

Within a year he had worked out the basic mechanism of the rainstorm at the heart of every shower; within another he had laid down the concept of the air mass that travels across the surface of the planet, carrying weather systems with it as it moves. Within a third year the Bergen School, as the group had become known, had crowned its achievements with Bjerknes's recognition that huge air masses operated on a

Above: The Norwegian, Vilhelm Bjerknes (1862–1951), was a pioneer of modern meteorological practice.

planetary scale, and in particular with his identification of the boundary that encircled the globe. Formed between the ice-cold polar air mass and the warm air mass from the subtropical latitudes, this boundary, more than any atmospheric feature, influenced the path of the cyclones that swept the northern hemisphere. In the shadow of the Great War from which they had all recently emerged, Bjerknes and his team identified the boundary as the front line in a battle between the cold air attacking towards the Equator and the warm air fighting back. He named it the "polar front." The cyclones that marked the sites of the most intense fighting were characterized by smaller "warm fronts" which revealed the path a storm was taking, and "cold fronts" that brought stormy showers of rain. It is a terminology born of terrible human conflict, but one which has remained with us to this day.

STORMS OVER NORWAY

The remarkable meteorological achievements of Vilhelm Bjerknes were kick-started by the privations of war. Norwegian agriculture was under immense pressure to boost production during the First World War, and Bjerknes argued that this was an opportunity for meteorologists to do their bit—and get some resources into the science. He supported the idea of providing a weather forecast service to farmers by telephone and was asked to set it up. To help do so he roped in the farmers themselves, as well as fishermen, lighthouse keepers and the Norwegian navy's watch stations, which were guarding against German submarines. In the end he found himself at the center of a network of seventy-five reporting stations across the country. Each delivered a weather report twice a day, providing the steady stream of data from which predictions could emerge. Bjerknes's first target was the understanding and better prediction of the low-pressure cyclones that bore down on Norway with relentless frequency, spinning anticlockwise as they moved in from the Atlantic. Until then meteorologists had assumed that the worst effects of the storms were to be found at the center of the low pressure and that by tracking that center it was possible to foretell the onset of the wind and rain. But within less than a year the data from Bjerknes's troupe of amateur weathermen revealed a very different and far more complex pattern.

Instead of forming a neat, circular shape, the area of precipitation looked more like a thick-bladed sickle. As it traveled east, for example, the sickle blade curved to the south-east while the handle stretched down towards the south-west. The inside edge of the sickle indicated the direction of movement, so he called it the "steering line;" the leading edge of the handle he dubbed the "squall line," because of the stormy rain that it seemed to bring. In practical terms this new interpretation provided a more accurate set of expectations for those on the receiving end of the weather. They now knew that when a gradually thickening mass of cloud, with rain, passed overhead as the steering line and "sickle blade;" it would be followed by an increase in temperature, and even a clearing of the skies, before a sudden onset of violent showers as the squall line swept by.

Although this was an accurate interpretation of the measurements, it dawned on Bjerknes the following year that the wedge of warmer air that was carried between the two cloud lines of the cyclone signaled something far more significant for meteorology as a whole. He realized that this was a moving mass of air which retained different characteristics from the air on either side. Suddenly it was no longer enough to explain the weather in terms of changing pressure and temperature: and the concept of the air mass was born. Bjerknes realized that the mechanism of the cyclonic storm was simply that as it tracked across the Earth the leading edge of the warm air at the steering line rose up over cold air in its path, and cooled and condensed to cloud and rain. And similarly, as the colder air of the squall line followed through it pushed under the warm air which had to rise suddenly, creating the violent showers.

It was then that he also spotted the blindingly obvious: the nature of the clouds that could be seen in the sky fitted perfectly with this three-dimensional model that he had worked out, and could therefore be a powerful indicator of the weather ahead. First, as the cyclone approached, high-altitude wisps of cirrus cloud would reach out ahead of its path; these would thicken to become cirrostratus as it drew nearer, thicken again to come lower in the sky to form altostratus, then come lower still to form the nimbostratus rain clouds of the steering line. After this there would be clear skies, but then the fluffy cumulus clouds of the squall line would appear, rising rapidly to thundery cumulonimbus as the cyclone passed on its way. Here for the first time was a scientific explanation for "reading" the clouds, an activity that had formed part of ancient weather lore for centuries (See Chapter One, p.18).

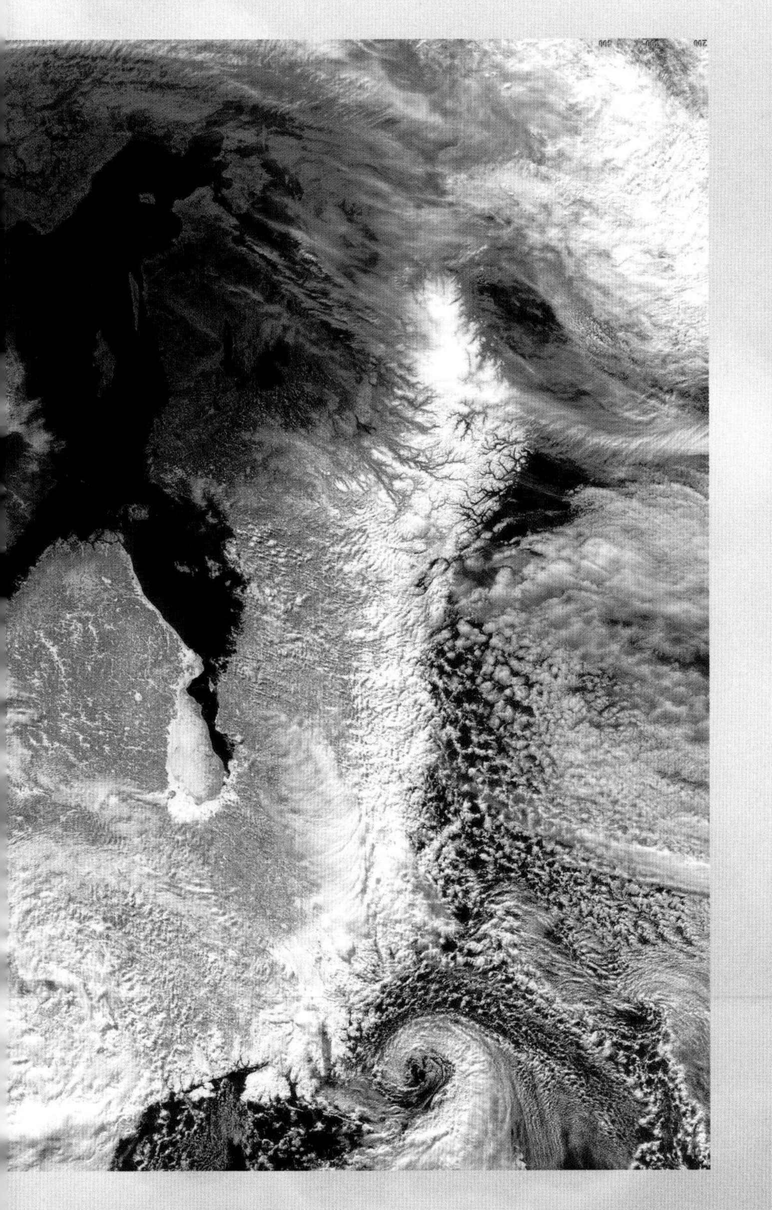

Far left: Amid the harsh glaciers of Norway, a network of farmers and fishermen provided daily observations to build the science of forecasting in the early twentieth century.

Left: Today's satellite images reveal that Vilhelm Bjerknes and the "Bergen School" of meteorologists were remarkably accurate in their description of the weather patterns over Norway almost a hundred years ago.

Above: The great flood of 1953 devastated the low-lying lands of Holland across the entire coast.

atlantic surge

Exposed to the full strength of the westerly Atlantic airflow, Great Britain and western Europe have long faced the dangers of a storm surge from the ocean. As well as driving the gales that lash the western shores, the winds have a direct effect on the ocean itself, an effect which is felt in places far removed from its source. The particular shape of the British Isles, and their short distance from the mainland European coast, have created the narrow channel of the North Sea. And at the sea's southern end the low-lying countryside of East Anglia in Britain and the reclaimed lands of the Dutch polders are vulnerable to events that begin far over to the west of Ireland. On the night of January 31, 1953 the greatest such storm surge in living memory struck those two regions with terrible consequences.

It began with an Atlantic storm that was devastating in its own right. In the north-west of Scotland hurricane-strength winds brought swathes of damage to the countryside, blowing down more trees than were normally felled by foresters in a whole year. A car ferry was sunk along the western coast, with the loss of 133 people. But what creates a surge is the perfect timing of the storm-force winds as they coincide with the natural movement of the tide, and that's what happened in 1953. An area of low pressure, just a typical depression, had formed in the north Atlantic south of Iceland, but on January 30, its pressure dropped further and the center of the depression began to turn eastwards. By early on the 31st it sat over the North Sea, between Scotland and Norway. This resulted in much stronger westerly winds flowing round the southern edge of the low pressure—and across the sea north of Scotland. These gales pushed against the surface of the sea, not only building huge waves, but also simply piling up the mass of water, the "surge," which added to the height of a tide that was already naturally very high. The surge crept around the north of Scotland and headed south. But as the depression moved away from the south of Iceland a ridge of high pressure had built up there, resulting in even stronger winds blowing southwards, along the western boundary of the depression. These force 10 and 11 winds drove the surge tide onwards to the south, down the eastern coast of Britain.

By the afternoon of the 31st, at Spurn Head in Yorkshire the shingle spit was breached; by early evening the seafront promenades of Mablethorpe in Lincolnshire were wrecked, and sea water began to flood the farmlands inshore. That night, the exposed coastline of north Norfolk bore the brunt of the oncoming water. Here the North Sea itself is shallower, so the effects of a surge are greatly amplified. Waves over 26 feet (8 meters) high battered the worn coastline, whose sea defenses have been constructed and reconstructed again and again since the sixteenth century. A ship was lifted up on to the quay at Wells-next-the-Sea; embankments were flooded and people began to drown. As the surge swept on round East Anglia, devastating the coastline of Suffolk, people had little warning of its severity—much of the network of telephone landlines further north along the coast had been put out of action by the high winds. Felixstowe and Harwich were flooded by midnight, and the surge swept on to Essex where many lost their lives on Canvey Island when the sea walls collapsed. They collapsed also along the banks of the Thames in the East End of London and caused widespread flooding, as what is estimated to be 22.6 million cubic feet (640,000 cubic meters) of river water swept into the single borough of West Ham.

A North Sea surge tends to follow a familiar path, down the east coast of Britain, then across to the Low Countries as it turns east along the continental coast before wheeling north up towards the coastline of Scandinavia, by which time it will have dissipated. In the Netherlands,

Above: Flood damage at Margate in Kent during the 1953 North Sea surge.

in the early hours of February 1, fifty dykes were breached and 1,800 people drowned.

Along the north Norfolk coastline that night, between King's Lynn and Hunstanton, where the surge reached its maximum height of almost 10 feet (3 meters) above the normal high tide, sixty-five people died. Many more might have done so, had the Norfolk coast not also been the scene of great bravery among the people who set out to rescue anyone they could from the floods. At the nearby US Air Force base at Sculthorpe a call went out for volunteers in the early evening, and a group with dinghies, aluminium boats and wetsuits arrived at a temporary headquarters at the Sandringham Hotel. The boats' propellers continually snagged in the debris and so the men took to wading through the water, towing the dinghies behind them, moving from house to house. One of the tallest men in the team was Corporal Reis Leeming, who throughout the night made journey after journey through the streets, sometimes up to his neck in water, sometimes sinking into ditches that appeared below him, battling to keep his boat stable in the high winds, to rescue a total of twenty-seven people, mostly women and children. Three times he waded the ½-mile- (1 kilometer-) length of the row of devastated bungalows behind the beach, calling out for survivors as he went, colliding with floating debris and even dead bodies. In the end he himself nearly perished when he collapsed, frozen, into unconsciousness in the water. But he was dragged to safety. Corporal Leeming's heroism was the more remarkable for his being a non-swimmer and he was awarded the George Medal in recognition of his bravery. Today the memory of the night, and of the families he was unable to reach, continues to haunt him and he is still overcome with emotion each time he recalls the events.

The 1953 storm surge was the largest on record that had flowed down the North Sea and the utter devastation that it caused prompted the creation of the Storm Tide Warning Service in Britain. It also set in motion plans to build the remarkable sea defenses that today line the Thames estuary and which culminated in one of the engineering triumphs of the last twenty

Opposite: The flooding of 1953 was the spur to building the Thames barrier which is known as one of the engineering wonders of the world.

years—the sleek and shiny Thames flood barrier, whose steel gates are designed to rise up out of the water to stem the flow of the highest storm surge that might be expected in a hundred years.

wettest place in europe

The journey of water around the planet can end in so many ways: as part of the salty Gulf Stream current that will sink once again to go round the thermohaline conveyor; as an Atlantic storm that will bear down on the west coast of Scotland or find its way into a storm surge down the North Sea; or it may simply stay aloft until the prevailing westerly winds bring it to the wettest city in Europe. Bergen, in Norway, is a small city with four times the annual rainfall of London. It receives 1,162,500 tonnes of rain each year; it rains on two days out of every three, and nursery-school children are dressed in special rain suits and hats so that they can be left to play outdoors all day. Its geography has placed it in the path of 80 percent of the low-pressure systems that emerge out of the North Atlantic and North Sea, and the mountains behind it form a barrier up which the moist air rises dramatically, rapidly condensing out as even more rain.

The people of Bergen are not really stoic—they rather enjoy their rain, and in October every year a rain festival is held when a procession of waterproofed merriment passes through the streets. Houses are designed with heated wardrobes for drying clothes, umbrellas come in different shapes and sizes as fashion accessories, and in a store-room in the city can be found a stockpile of cans of "Bergen Rain," packaged specially as souvenirs for the tourist trade. In 1938 the population found itself glued to the weather forecasts on the radio when it seemed that the rainfall record was about to be broken, for they are proud of their accolade as the wettest city in Europe.

It is perhaps fitting that this remarkably damp city should also have been the home of the Bergen School founded by Vilhelm Bjerknes (see p.117), who is rightly credited as being a founding father of modern meteorology. And it is fitting, also, that we have chosen here to end our water's journey, from one side of the planet to the other, with rain cascading over the waterfalls and rivulets that will take it from the Norwegian fjords back to the ocean, to form one of the many outflows that eventually join the deep Atlantic water, and to set off again on another journey around the globe.

Above: Rain is part of everyday street life in Europe's wettest city: Bergen in Norway.
Opposite: Spring meltwater swells the River Ranma in Norway.

Of all the elements, cold remains our deadliest enemy. Every year snow falls on half of the land surface of the northern hemisphere, and ice covers much of the northern oceans. In their many forms—frost, snow, blizzards and avalanches—the weather's icy fingers reach far out from the poles across the people of the globe, even to touch the Equator itself. And with the cold comes death.

chapter four

cold

a cold planet

Ever since the first astronauts looked down on the Earth as they circled the globe in fragile spacecraft we have come to know our world as the blue planet. That iconic image of the tiny Earth floating in a black void, taken by Apollo spacemen half-way to the moon, is dominated by the blue of the ocean. And in 1990 the Voyager space probe took a remarkable photograph of the entire solar system from its vantage point at the edge of deep space, with the planets so small that they could hardly be seen. In that image, too, the single pixel of light that represents our world is blue. We live, as Carl Sagan put it then, on a pale blue dot.

But view the planet from above the North Pole at the close of the Arctic winter and a very different impression emerges: the Earth appears largely white. Every year snow falls on half of the land surface of the northern hemisphere, and ice covers much of the northern oceans. Some of that snow and ice never leaves, and the power that is locked up inside it is perhaps the most terrible of any on the planet. For the accumulation of frozen water in the Arctic keeps it cold, holding the temperature of the air above it down as far as -58°F (-50 °C). Its icy fingers each year reach far south across the peoples of the northern hemisphere, even to touch the Equator itself. The Arctic is the home of winter. And with the cold comes death. Human existence has always been characterized by a war against the cold in its many forms—frost, snow, blizzards and avalanches. Cold is our most lethal and enduring enemy.

Right: Even at its smallest extent, at the height of summer, the ice of Greenland and the Arctic is a powerful presence in the northern hemisphere.

SUMMER AND WINTER

The mechanism is simple. Over a period of 365 days, our year, the Earth quietly orbits around the sun, traveling through space at a rate of somewhere over 1½ million miles (2½ million kilometers) a day. It also spins daily on its axis. This axis is roughly perpendicular to the sun, so the sunlight strikes the Equatorial regions directly and most intensely, but shines on the polar regions obliquely, spread over a wider area and barely glancing against the poles themselves. The solar radiation—light and heat—that arrives from the sun is called "insolation" and, because of the oblique angle at

Above: 'Midnight sun': the polar sky in northern Norway.

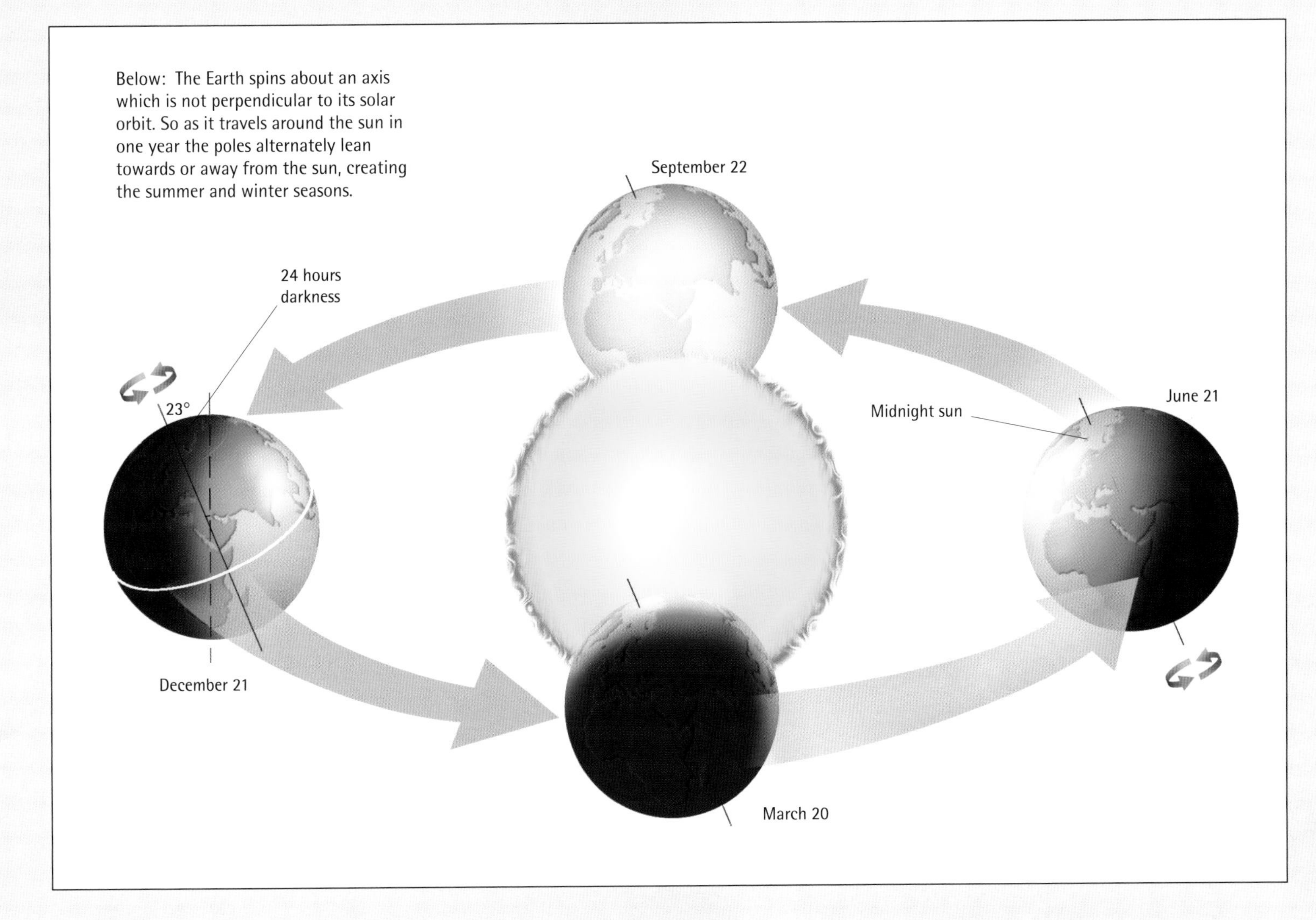

Below: The Earth spins about an axis which is not perpendicular to its solar orbit. So as it travels around the sun in one year the poles alternately lean towards or away from the sun, creating the summer and winter seasons.

which it strikes the polar regions, it has to pass through a much greater thickness of atmosphere on its journey to the surface of the globe. As it passes through the atmosphere it is reflected and scattered by air molecules, particles of dust, water droplets in cloud and atmospheric gases. So, by definition,
a lot less heat strikes an area at the poles than a similar area at the Equator. The result is a world with ice at either pole. But then it gets more complicated. The Earth's axis is not quite perpendicular to the sun. It tilts at an angle of 23.5 degrees but, as it always tilts at this angle, the North and South Poles take turns in leaning towards the sun as the planet orbits the star. When the North Pole is leaning towards the sun the Arctic has its brief months of warmth. On March 20 the sun rises above the horizon at the North Pole and, although it remains low in the sky throughout the Arctic summer, it never sinks below it again for a further six months. At the height of summer, June 21, the entire Arctic Circle—everything north of latitude 66.5 degrees—experiences twenty-four hours of daylight. This earns it its title "land of the midnight sun," and during this period, further south the people of the northern hemisphere experience their summer. When the pole is leaning furthest away from the sun no solar radiation falls on the planet's surface inside the Arctic Circle, the region is plunged into total darkness, temperatures fall and winter takes hold of the north. The moment of maximum tilt away from the sun arrives on December 21 each year. That, officially, is the astronomical start of the northern winter, although few people will not have felt its impending arrival for many weeks.

greenland ice

Above: The coastline around the southern tip of Greenland is fragmented into many inlets and fjords.
Right: The Kangerdlugssuaq Glacier in Greenland, at a point where two rivers of ice join together and flow towards the sea.

The most striking Arctic land-mass in a polar view of the planet is Greenland. And what dominates Greenland is ice. The Greenland ice sheet covers over 650,000 square miles (1.7 million square kilometers), and, after Antarctica, it is the largest permanent mass of ice on the globe. The bedrock of Greenland itself lies close to sea level across most of the land, but the statistics of the ice that lies above it are so enormous that it is hard to make them do justice to its awe-inspiring scale. The total volume of the ice is 670,000 cubic miles (2.8 million cubic kilometers)—8.4 per cent of the world's total—and its average thickness is 1 mile (1,500 meters), rising to 2 miles (3,200 meters) of pure ice in the southern dome that forms the harsh landscape. But the ice sheet is far from a solid mass. Instead it accumulates with falling snow and ice in the central higher regions. Like a conveyor belt, it moves constantly, and is drained by the flow of glaciers that slowly carry huge volumes of ice to the edge of the ocean and tilt them thunderously into the sea, a process known as the "calving" of icebergs. The fastest-moving glacier in Greenland, known as Jakobshavns Isbrae, lies on the western side. It drains over 6.5 per cent of the Greenland ice each year and pours out at a rate of up to 4 miles (7 kilometers) a year. And, if that is not impressive enough, if the Greenland ice sheet, along with the other smaller ice caps that are dotted across the Arctic, were to melt, the global seas would rise by 20 feet (6 meters). Unquestionably, Greenland is the largest inhabited ice cube on Earth.

adapting to ice

The people who live in Greenland have developed a remarkable ability to survive. Known for many years as Eskimos, which is actually a Canadian Indian word for "eaters of raw meat," they are more generally called "Inuit," which means the "real people." Some 45,000 of them inhabit Greenland today—about half of them live around the Arctic Circle. They have adapted to this most inhospitable of climates by developing real ingenuity in the practicalities of life. Traditionally there were few settled communities: some clusters of houses made from stone; some shoreline fishing settlements made from driftwood; but mostly the people always traveled. Their focus on hunting marine mammals—the walrus, seal and whale—has required them to move great distances, carrying their means of survival with them. Their ability to make tools and equipment from whatever material is at hand is legendary. Sealskin clothing, harpoons of walrus ivory or whalebone or antler, stone-bladed knives, wooden dog-sleds with runners made from frozen lengths of dried fish, and the famous igloo made with snow—all these are the signs of a culture closely in tune with its environment. Today the dominance of the European and American cultures has brought the Inuit a more settled existence in simple contemporary buildings, but the values they aspire to are still courage and hardiness. Their awareness of the fragile balance between their own survival and that of the creatures they live alongside is revealed in their need to appease the souls of the animals they hunt. Physically the Inuit have evolved traits that equip them well for their environment: their short, squat bodily appearance results in a particularly low surface area to volume ratio, thus allowing their bodies to conserve more heat, which is otherwise lost from the surface of exposed skin. Their vascular system is also able to maintain continuous circulation of the blood in their extremities—fingers and toes—at temperatures below the point at which the rest of us would begin to suffer frostbite. Adaptive traits like these are actually achievable by anyone who adopts a careful regime of prolonged exposure to the cold, but they are characteristics that the Inuit's thousands of years of extreme existence have brought to perfection.

Above: The light inside an igloo reveals its structure.
Opposite: An Inuit fits a sealskin line to a harpoon.

the source

Strangely, the weather in the heart of the Arctic ice-cap can be remarkably unchanging. The large expanse of frozen surface offers few alterations in temperature to the air mass that floats above it, so few changes occur in the density of the air. As a result there are few differences in pressure, and thus little wind. The air is slightly warmer than the frozen waste below, so the lower layers of air are kept chilled by contact with the ice. This condition, with warm air resting above colder air, results in a very stable air mass, with little reason for air to rise or fall. Vertical mixing

of air normally occurs because temperature increases close to the earth cause the warmed air to rise through colder air above, but here there is nothing to make the lower air rise. The huge mass of stable air means that any dust or pollution simply hangs in the sky, to create a gentle haze, before sinking slowly to the surface of the ice. (This gathering of airborne pollutants is something that has turned out to be very useful for scientists trying to measure the effects of human activity on climate change, as will be explained in Chapter Six.) But over much of the high interior of Greenland the air is clear, and the ice sheet is slowly added to by tiny ice crystals that form in the cold air and settle as a fine "diamond dust" on the ice below. This stable quality of the air in winter over Greenland and the Arctic ice-cap is shared with other large areas of the Earth's surface. These include the subtropical oceans and desert regions in summer—although for very different reasons of temperature, pressure and humidity. All the areas share the characteristics of being large expanses of flat, uniform surfaces, with virtually no variation in temperature and pressure across them, and so only light surface winds. They are known as "source areas," for, while they themselves are remarkably stable, they provide the basis for weather patterns that emerge to violent effect elsewhere on the globe.

But round the edges of the land mass of Greenland, cyclonic storms do build up and work their way along the coast, occasionally pushing inland to create the snowstorms that fall on the ice cap. Because the movement of air across the interior is so restricted, the storms can move very quickly up the coast, and some of the most rapid drops in air pressure recorded anywhere occur around Greenland. This means that these coastal storms are of a very different nature from the stable air mass inland. They can be some of the most violent storms experienced anywhere, other than hurricanes themselves.

weather war

Greenland's dominant position in the Arctic has meant that its presence is a powerful influence on the northern European, North American and North Atlantic weather. So much so that it has built up a reputation for being the barometer of our climate. Know the weather around Greenland, and you have a pretty good idea as to what is on its way further south. Indeed this rule was at the heart of a little-known battle that was played out during the Second World War—it became known as the "weather war."

The outcome of battles on land and at sea throughout history has so often been decided by the weather: the rains at Agincourt left a sea of mud in which the heavily armored French knights became entrapped as living targets for English longbowmen. The gales of the North Sea finished off the destruction of the Spanish Armada, and ended the hopes of Philip of Spain to be king of England. The army of Napoleon was consumed by the snows of the Russian winter; and even as far back as the time of ancient Rome the extraordinary force of men and elephants that Hannibal led over the Alps was decimated by the mountain snowstorms—although he, of course, went on to win. But by the time of the Second World War a new kind of force was rising to prominence: war in the air. The influence of weather on the course of airborne warfare was both critical and often fast-changing, and as a result the military leadership on both sides of the conflict placed great store on their ability to predict the nature of the elements. There is little doubt that the Battle of Britain could only have been fought with the intensity that it was, and arguably could only have been won, because of the British weather conditions in the warm summer of 1940.

Yet in the 1940s there were of course no weather satellites to reveal Atlantic storms and cold fronts building up to sweep down into Europe, and weather prediction was as much an art as a science. The meteorologists needed to gather data concerning impending storm fronts from as close to their source as possible, which meant Arctic Norway, the Norwegian Sea and Greenland, but with good information from these regions reasonably accurate forecasts,

Above: Napoleon's army burned its standards for warmth during the fateful Russian campaign of 1812.

up to forty-eight hours ahead, could be prepared for military commanders.

In April 1940 Germany invaded Denmark and Norway, and immediately the Allied powers feared that their enemy could gain control of data from the weather stations that those two countries managed in the Arctic. Until then all their data had been transmitted by radio in an open fashion, using internationally known codes, but after the invasion the weather men began to use codes known only to the Allies. Greenland was never actually occupied by Germany, but the Germans realized that they would physically have to get control of the weather stations. From that

moment on the Norwegian island of Spitzbergen, and the whole of Greenland, became embroiled in a game of cat and mouse until the end of the war. The Germans' first attempts to control the weather forecasting from Greenland met with disaster. A four-man meteorological team that arrived on an old whaling ship was quickly captured, and several other ships were intercepted by the British navy, which still controlled the Arctic waters. But the Germans eventually succeeded with a weather ship that transmitted weather reports from around Iceland for seventy-six days before being captured, and with specially converted Junkers 88 and Heinkel 111 bombers that flew out of occupied Norway.

Meanwhile the British landed a weather team at the weather station on Jan Mayen Island in the Norwegian Sea. The team was stranded when a storm sank their ship, and had to abandon the station, but over the course of the autumn of 1940 all of the eastern Greenland weather stations were secured into British hands. After one of the worst winters on record, when neither side could make any gains, the British were back on Jan Mayen Island in spring 1941, only to be pounded by a German air attack. In May the British were still patrolling the Greenland seas and captured a weather ship with its famous Enigma code-machine intact, together with all the German shipping forecast settings. The more well-known capture of the U-boat *U-110* happened only two days later, and together these successes helped the British break the German naval codes, to dramatic effect throughout the remaining period of the war.

The Germans were not be outdone, however, and had set up a weather station on Spitzbergen. The British captured this in August 1941 and successfully held off a counterattack for three days by transmitting fake reports suggesting that the cloud base was too low to allow any aircraft to fly. But after the British had pulled out—together with the entire population of the island—the Germans landed a ten-strong team of weather men who were successfully transmitting reports by the onset of winter. They in turn were

Above: A soldier of the Sirius sledge patrol.

hounded by a party of free-Norwegian ski-troops who were themselves attacked by German fighter-bombers. The Germans pulled out finally in July 1942, in the face of an onslaught by two battlecruisers, four destroyers and a battalion of infantry. But the weather data kept coming to Germany because they had left behind an automated weather transmitter which the Allies never found. And in August 1942 two German weather parties and their equipment were secretly landed in eastern Greenland, where they succeeded in surviving undetected—indeed it was an achievement to survive at all—throughout the brutal Arctic winter. However, the Danish army in Greenland had set up a special sled-patrol with the help of local hunters and it was tasked with patrolling the coasts to detect any German activity. In March 1943 a small dog-sled team ran into the twenty-seven-strong German weather unit and were badly defeated, although one man managed to escape and make a heroic 600-mile (1,000-kilometer) trek back across the ice to reveal the Germans' position. The resulting counter-raid by American B-24 bombers wiped out the German party.

By the middle of 1943 Hitler himself personally intervened to order the reinvasion of Spitzbergen, in a massive operation code-named "Zitronella," that involved the battleships *Tirpitz* and *Scharnhorst*. The Germans succeeded in reoccupying the island after a mere four hours but, in the face of Allied naval supremacy in the region, they knew that they could not hold on to the island. They withdrew after a day, and the Allied weather station was transmitting again a month later. But, under cover of the impressive activity of the two iconic German ships, a trawler deposited another secret team on Greenland who again survived undetected, until the middle of 1944 when the patrols caught up with them once more. Again the German weather men left an automated station behind them when they pulled out, but that was the last of the weather signals that the Axis powers succeeded in getting from the ice cap of the north. The Arctic weather war may have been fought in one of the most remote and inhospitable parts of the world but, by the time it was over, it had involved eighty-nine ships, six aircraft squadrons and fourteen companies of infantry.

The secret weather reports from Greenland were fed into the forecast that dictated the Allies' timing of the D-Day landings in Normandy. As a testament to the importance of the weather war, the last Axis force to surrender to the Allies at the end of the conflict were members of a German weather unit, several months later than the rest of their compatriots, on September 9, 1945.

Above: Robert Edwin Peary (1856–1920), who became a victim of an arctic mirage during his search for the North Pole.

survival and wonder on ice

The North Greenland Sledge Patrol, which fulfilled such a vital role in policing the ice sheet during the war, was disbanded in 1945. However, five years later it re-emerged secretly as part of a cold-war strategy to maintain the security of the Arctic, given the very close proximity of the Soviet lands on the other side of the pole. The new name for the patrol was "Sirius," after the star that formed the brightest point in the "great dog" constellation. Today the teams are still on patrol for eight months of the year, tasked with surveillance of the 1,300 miles (2,100 kilometers) of uninhabited coastline in north and north-east Greenland. Only twenty-seven people stay in north-east Greenland over the winter, for physically and mentally this is one of the most arduous military missions in operation anywhere. A two-man team can be traveling alone with its dogs for up to four months, with a tent as the only protection against the snowstorms that roll in from the coast. The men experience extreme low temperatures and hurricane-strength

Above: This non-existent "iceberg" is in fact a Fata Morgana, a mirage over the arctic ocean.

blizzards. The physical dangers are intense—the surface of the eye can be frozen by an icy wind that blows directly on to the face. Cold impairs the function of nerves, decreasing sensation and reducing manual dexterity. If bare skin touches metal it will rapidly freeze to it, because the metal quickly conducts all the heat away. At -58°F (-50°C) bare skin itself will freeze within a minute. And unprotected frozen skin will result in frostbite. When human tissue freezes ice crystals form in the fluids that surround the cells of the body. As a result the concentration of chemicals in the fluids becomes stronger (because the water has turned to ice), and so more water is drawn out of the cells themselves to compensate. This means that the cells shrink in size and concentrations of salt inside rise to a critical level, damaging proteins and ultimately killing the cell. Deep frostbite is a one-way process.

But the perils of the two-year stint of duty on the sledge patrol are to a degree compensated for by the opportunity to witness some of the most remarkable products of weather phenomena. The temperature inversion

(the still, cold air held against the Earth's surface with the warm layer of air above) that lies over the bulk of the interior allows sound to be heard over phenomenal distances. The light reflected from the vastness of the white surface is so great that planes can fly by the light of a half-moon alone. In sunshine the light is so bright that shadows are almost entirely banished—it is hard to see undulations or crevasses—and the surface and sky blend into a continuum where it is impossible to judge distance. Light rays passing through the temperature inversion at low angles are also bent, creating the strange effect of Arctic mirages, where objects that are in fact below the horizon are seen as being above it.

One of the most famous, and embarrassing, of these mirages was "Crocker Land." In 1909, the North Pole was first reached by Robert E. Peary, an American explorer. But three years earlier he had been surveying the northernmost part of North America when, staring north through binoculars, he had seen a range of snow-peaked mountains in the far distance. He believed it to be a part of the northern landscape that would lead to the pole and he named it Crocker Land, after one of the financiers who supported his expeditions. Funding was raised some years later for an expedition to explore Crocker Land, and it was duly spotted far to the north. But on arriving where it should have begun, the explorers found that they were still on the polar ice-cap, with the "mountains" still to the north—and that they had been tempted by one of the most expensive mirages in history.

snowflakes

Wherever you are in the world there is almost always snow somewhere above you. In the tropics, or in midsummer, it may be confined to wisps of ice crystals in the highest cirrus clouds, where temperatures are constantly well below freezing. But over the ice sheets of the Arctic and Antarctic ice crystals can form in the air even very close to the ground, and snow reaches down to the surface. In fact, almost all the precipitation that falls on the planet begins its life as crystals of ice, whether they end up as snow on the ground or melt to become rain on the journey down.

seeds of snow

Snowflakes are collections of ice crystals, loosely bound together into a sort of puffball, that can grow to surprisingly large sizes. The flakes form in the parts of clouds that are filled with moisture and where the temperatures are at freezing or below. But what they need to start with are "seeds." Floating around in the atmosphere up to heights of 9 miles (15 kilometers), are tiny particles: they could be ash from a volcanic eruption in the Philippines, pollen thrown up by a gust of wind from a meadow, dust from a desert storm or even minuscule fragments of a meteorite that burned up in the atmosphere. Whatever the seeds are, molecules of water collect on these tiny particles and an ice crystal forms. By a process called vapor pressure, which draws more water vapor towards the ice crystals, they grow to become snow crystals. They also go on growing by colliding with other "supercooled" water droplets which freeze to them, and by colliding with other ice crystals which fracture to form new "seeds," so that a chain reaction occurs with more and more ice crystals building up in the cloud. Eventually the growing crystals become heavy enough to counteract the updrafts of air in the cloud and begin to descend, with the pull of gravity at a gentle 2 to 3 miles (3 to 5 kilometers) an hour—even as fast as 9 miles (15 kilometers) an hour—partially melting and becoming sticky as they go. As they fall the collisions continue and the crystals stick together, forming snowflakes.

SNOWFLAKE STARS

No two snowflakes can ever be the same, but they can be classified into six broad types of crystal: needles, columns, plates, columns capped with plates, dendrites and stars. Exactly what kind of snowflake drops from the sky to land on our heads depends on atmospheric conditions such as humidity, temperature and wind. Snowflakes generally remain small when temperatures are well below freezing, but they can grow very large when the air warms and becomes closer to freezing point, because the snow becomes relatively wet and the flakes stick together. Some of the largest snowflakes ever recorded—which fell over England in April 1951—were 5 inches (12.5 centimeters) across.

High in the upper atmosphere, where temperatures fall below -4°F (-20°C), crystals are tiny, held aloft by the updrafts and bouncing off each other because they are very cold. The ice forms crystals shaped like bullets, needles or columns in the dry air. They are associated with the optical phenomenon known as the "sundog," a halo of orange-and-green light that appears around the sun because its rays are refracted by the crystals of snow.

The familiar, star-shaped crystals form at between 23°F and 14°F (-5°C and -10°C) and these are large enough to fall through the air at a gentle about 1 or 2 miles (2 or 3 kilometers) an hour. But these tiny crystals are fragile and many break during collisions with other crystals or with supercooled water droplets, so barely a quarter of those reaching the ground, whether as individual crystals or as collections making up snowflakes, arrive with their beautiful geometrical structure intact.

STRUCTURE OF SNOWFLAKES

The philosopher and mathematician René Descartes was the first scientist to describe the real structure of a snow crystal reasonably accurately, based on what he could see with the naked eye. But the classic image of the snowflake, the beautiful pattern of a silvery-white star or hexagonal lattice which now so often adorns the tops of our Christmas trees, was captured by Wilson A. Bentley and published in the 1930s. During his lifetime Bentley, an American farmer with a passion for microphotography, produced some five thousand snow-crystal pictures, recording the sheer beauty of shapes that look more like perfect engineering designs than constructs of nature. The structure of the crystals reveals the dominant hexagonal shape, which stems from the fundamental molecular structure of water. The snowflake has crystal branches sticking out of each side at an angle of 60 degrees. The angle between these branches is therefore 120 degrees, and it is no coincidence that the angle between the two hydrogen atoms that attach to an atom of oxygen to make up a molecule of water is also 120 degrees.

Bentley's book inspired Ukichiro Nakaya, a Japanese physicist, to begin a painstaking program of research into the structure of snow crystals; over many years he succeeded in photographing many of the different varieties of snow crystal and eventually managed to re-create them in the laboratory. It was Nakaya who worked out the relationship

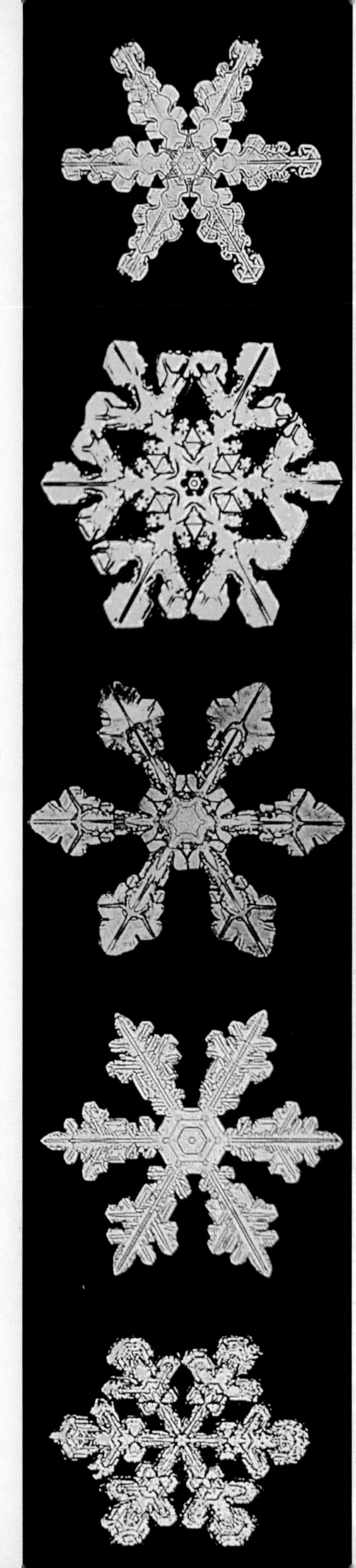

Left: Microscopic images of snowflakes, from the collection of photomicrographs by Wilson A. Bentley. Above: A "sundog" seen over Baffin Island, Canada. This halo around the sun is created by ice crystals in the atmosphere which refract the sunlight.

between the shapes of snow crystals and the atmospheric conditions in which they were created, and summarized his work in the form of a chart, known as the "Nakaya diagram." From this it is possible to "read" the meteorological information "written" on a snow crystal, because we can infer the weather conditions in the upper air by observing snow-crystal shapes on the ground. Nakaya often referred to the snowflake as "a letter from the sky."

SOUND OF SNOWFLAKES

One of the features of a winter scene, with its familiar blanket of snow, is the apparent stillness and quiet of the surrounding air. But snow doesn't necessarily mean silence. Recently scientists have discovered that it has its own sound. The lattice structure of a snowflake results in a significant quantity of air being trapped in the ice. In fact, a snowflake consists of only about 10 percent water molecules. When a snowflake lands on water the ice melts and a bubble of air becomes caught within the water. The theory is that this tiny bubble then immediately works its way to the surface, but before it gets there and bursts, the surface tension and pressure of the water cause the bubble to pulsate. These oscillations cause a very high-pitched screech at a frequency of between 50 and 200 kilohertz, far higher than the human ear can hear. An individual snowflake thus emits its own scream, which lasts barely a ten-thousandth of a second. The screech of a snowflake was first detected in the 1980s by researchers who were monitoring the sound that rainfall makes on water, in order to establish how it might interfere with the sonar-detection systems of submarines.

Left: A computer-generated super close-up image of a snowflake reveals its detailed crystalline structure.

the lost squadron

The snow that falls on the Greenland ice is among the most beautiful works of nature that can be seen—magnified under today's high-powered microscopes the crystals are like perfect engineering structures (see pp.144–5). But they are also the guardians of a remarkable history. The layers of snow that fall mark the beginning of a conveyor belt that ends with the glaciers pouring ice into the Arctic Ocean, and this process takes at least a hundred thousand years to complete. Every year the layers of snow are gently compacted down by the following year's fall. Over the millennia the snow crystals are compacted further to form rounded grains of ice, which fuse together to become ice crystals the size of footballs. Gradually the snow pack becomes even denser, to become glacier ice, heavy enough to sink and to flow downhill. So the ice sheet does not build up its thickness; rather, the layers of snow gradually sink to the bottom where they are carried out by the flow of the glacier. A striking illustration of this process is thrown up by the remarkable story of the lost squadron.

By the summer of 1942 the shortage of warplanes stationed in Britain was very acute and America's military leaders initiated Operation Bolero in an effort to increase the number of aircraft they could supply. Shipping was considered too risky because of the losses the North Atlantic convoys were incurring at the hands of the German U-boats. So it was decided

Above: The edge of the Greenland ice sheet, spreading out into the sea. To the right, an iceberg has recently been calved and is floating away.

Above: The remains of one of the "lost squadron," entombed in the Greenland Ice.

to fly some of the aircraft directly, refueling in Labrador, Greenland and Iceland before reaching Britain. On July 15 twenty-five US Army Air Corps crewmen were on just such a mission, delivering two B-17 Flying Fortress bombers and six P-38 Lightning fighters, when they ran into appalling weather over Greenland. The result was that they had to make a forced landing on the ice cap, where the aircraft were left for the weather to take its toll. And quietly, the planes disappeared, snowed on, year after year, until they gradually became part of the ice cap. After forty-six years a team of enthusiasts attempted to locate the aircraft and recover them. Surveys of the ice and records of snow patterns led the searchers to calculate that the aircraft would have reached some 50 feet (15 meters) below the surface, but when ground-penetrating radar finally pinpointed their location and the remains had been uncovered, they were found to be under 260 feet (80 meters) of glacier ice. Half a century of glacial movement downwards had not served the aircraft well, for an attempt to dig out one of the Flying Fortresses revealed that it had suffered serious structural damage. A later attempt to recover one of the Lightning P-38s was successful as the smaller planes had proved more resistant to the crushing forces within the ice. This aircraft has now been returned to the USA, where it is undergoing restoration. The lost squadron has proved to be of interest not only to aviation enthusiasts. It provides remarkable evidence for the rate at which ice accumulates on the Greenland ice sheet, which is of current critical interest to scientists trying to find a measure of the effect of climate change in different parts of the globe.

polar surge

The view from a satellite over the North Pole shows the clouds that are carried by the polar jet stream as the winds swirl their way from west to east around the rim of the Arctic, fluctuating to the north and south with the shifting pressures of the polar front below. But trapped inside that loop is also the clear, cold air that hangs over the Arctic ice. This huge volume of essentially stable air is one of the best examples of a phenomenon that, it has recently been realized, plays a critical role in dictating the weather experienced further to the south.

It is what is known as an air mass. With the long hours of winter darkness, and the brutal cooling effect of the ice below, this air becomes progressively colder and colder and very stable, forming a huge high-pressure dome over the region until a change in the jet stream can trigger part of it to slide southwards. Over the flat, open lands to the north of the USA there is little to stop this movement, and what develops is known locally as the "Siberian express" or, more correctly, as a polar surge.

This blast of icy air is continental in size: over the USA, for example, its fingers can reach as far south as Texas and Florida, rapidly chilling the air and ruining crops overnight. In one well-documented outbreak, over the winter of 1983–4, an Arctic air mass brought with it winds of up to 45 knots and some of the lowest temperatures ever recorded in America, including an unbeaten record of -65°F (-54°C) in Utah. Over a period of almost two months up to 90 per cent of the USA was covered by the polar air.

It was a polar surge, too, that lay at the heart of the extreme weather at Cape Canaveral on the morning of January 28, 1986, when NASA went ahead with the launch of the space shuttle Challenger, despite fears from the manufacturers of its solid rocket boosters that the now-infamous "O"-rings would not perform their essential function of sealing potential fuel leaks when they were so cold. Tragically the engineers were proved right, as the world saw on its television screens later that day when the shuttle exploded.

the world's worst-weather station

The Abenaki Indians named it "Agiocochook"—the Home of the Great Spirit—but at an altitude of barely 6,200 feet (1,900 meters), and set among many smaller peaks in the aptly named White Mountains of New Hampshire, Mount Washington does not at first seem a likely candidate for the accolade of having the world's worst weather. Yet the band of meteorological

Above: Large quantities of ice coated the launch pad on the morning of the fateful launch of the space shuttle Challenger.
Opposite: The launch of Challenger, January 28, 1986.

volunteers who man the weekly shifts at the Mount Washington observatory are in little doubt. The key is the lonely mountain's position. Although dwarfed by mountains such as those found in the Alps or Himalayas, Mount Washington is very exposed, and the summit stands like an island, directly in the path of winds blowing from several directions. In particular a wide valley opens towards the south-east directly to the coast, so winds from the Atlantic Ocean are funnelled, unchallenged, towards the weather station, accelerating dramatically as they race up the steep sides towards the top of the mountain. On April 12, 1934 the wind was blowing precisely from the direction of the valley and the ocean beyond and it built to a world record wind-speed measurement of 231 miles (372 kilometers) per hour. This record has now stood for almost seventy years, as has the wooden-frame building from which the measurements were taken, today still chained to the ground to prevent it from being blown away. (A measurement of 236 miles (380 kilometers) per hour over Guam during Typhoon Paka in 1997 was briefly hailed as the record-holder, but was later disqualified because it was believed that the performance of the anemometer taking the measurement had been distorted by the combination of high wind and heavy rain that struck it.)

The mountain also lies in the frontal zone where the air masses from the Arctic and the tropics meet, and is at the confluence of the three main streams of cyclones that bear upon the north-east USA. As a result, conditions for a tour of duty on Mount Washington are truly abysmal. The average annual temperature is 27°F (–30C); the minimum ever recorded was -47°F (–44°C), while the warmest it has ever been was

Above: Rime ice on the branch of a tree.
Left: Mount Washington, in the White Mountains of New Hampshire.

still only a brisk 72°F (22°C). For two and a half months of the year wind speeds of over 100 miles (160 kilometers) per hour are experienced, it is foggy for three hundred days and every month sees at least some fall of snow. Indeed, to say that there is snowfall is misleading. The truth is that the snow generally blows sideways in the icy winds, whose harshness causes the snowflakes to condense into pellets and globules that can do far worse than simply sting as they strike the skin. One particularly common feature is rime ice, where supercooled droplets of water, making up the fog shrouding the mountain or falling as rain, freeze immediately on contact with a solid surface. Much of the daily chore of the weather crew consists of breaking the rime ice off the meteorological instruments, where it builds into huge, feather-type structures, like horizontal icicles, blown out by the winds and increasing at a rate of up to 12 inches (30 centimeters) per hour. On some days the skin can freeze within thirty seconds, goggles have to be worn at all times, chunks of flying ice can cause serious injury, winds can hurl an observer to the ground and visibility can vanish in an instant.

The meteorologists who brave these conditions are local celebrities. They provide regular weather forecasts live from the summit on morning radio programs and recount tales of their attempts to stand upright in the gales when they step outside every hour to make regular weather observations. One tradition among them is the "breakfast of champions" where an individual is challenged to sit out at a table and take breakfast in a 93-mile (150-kilometer) wind: cereal poured from a box flies off before it reaches the bowl; milk and sugar go the same way; hot coffee blows away horizontally; spilt orange juice freezes before reaching the ground; crockery, cutlery and even the table disappear into the wind.

But the Mount Washington observatory has a very serious purpose. In addition to the regular monitoring of the weather, it is also home to a variety of scientific research projects aimed at understanding some of the many mysteries of meteorology. It has been an ideal location to test laser-radar for measuring winds at altitudes up to 7½ miles (12 kilometers) above the mountain, a technique that meteorologists hope will improve the accuracy of forecasting as high-altitude wind data are currently sparsely recorded throughout the world. It is also a perfect location to study the formation of rime ice itself, which is a considerable danger to aircraft that find themselves flying in extreme weather. And in addition to the meteorological research, the appalling weather makes Mount Washington the ideal training site for climbers who intend to tackle the great mountains of our planet. In fact, the winter worst of Mount Washington can be as bad as that encountered on Mount Everest.

superstorm of '93 –new york

On March 13, 1993 the winds at the Mount Washington weather observatory in New Hampshire gusted to an extraordinary 145 miles (233 kilometers) per hour, and over the following two days record low temperatures for that time of year of -24°F (–31°C), were recorded. On the same day Jeff Smock and Bill Simmons were working on the renovation of Smock's converted barn in upstate New York. It had begun snowing in the afternoon, but that was nothing unusual for the region in March, and at around 5:30pm the two men drove to town to get some supplies. On the return journey their vehicle got stuck in snow just out of town but, well used to an east-coast winter, they decided to walk the three or so miles on to the house. It was a decision that they came to regret, for Superstorm '93 had arrived in their county.

By the time the superstorm subsided it had left over a meter of snow across the states of the eastern seaboard, with drifts up to 10 meters deep ●

By the time it subsided it had left over 3 feet (1 meter) of snow across New York and the other states of the eastern seaboard, with drifts up to 30 feet (10 meters) deep. Every major airport on the east coast had closed, half the total population of the USA had been affected and over 270 people had died.

It had begun in the early hours of March 12 with what the weather men call a "disorganized area of low pressure" in the Gulf of Mexico just south of Texas. At the same time, at very high altitude, the polar jet stream was taking a long loop south along the line of the Rockies and turning east over Texas. In its wake a huge cold air mass was slipping down from the Arctic, creating an icy region of high pressure covering much of the northern USA. The north-westerly winds meant that the Arctic air was transported further and further towards the south and east, bringing with it the conditions for snow at low levels. The low pressure over the gulf and the shift of the jet stream would themselves have been enough to create a storm, but there was a third factor as well: a cyclone moving east of Nova Scotia and the returning jet stream flowing at over 125 miles (200 kilometers) an hour were converging to create a "confluence" of high-level air over the eastern USA. This was damming up the cold air over the continent, allowing a succession of intense fronts to form around its perimeter.

By noon on the first day, the low pressure had consolidated over the Gulf of Mexico and thunderstorms could be seen developing above Texas on the satellite images that began increasingly to worry weather forecasters across the USA. On the ground, ice pellets and snow began to fall as the storms met the cold air rushing south. The cyclone began to move progressively eastwards towards Florida, with atmospheric pressure dropping to record lows. As the afternoon wore on cumulus cloud built up and a "squall line" of thunderstorms developed

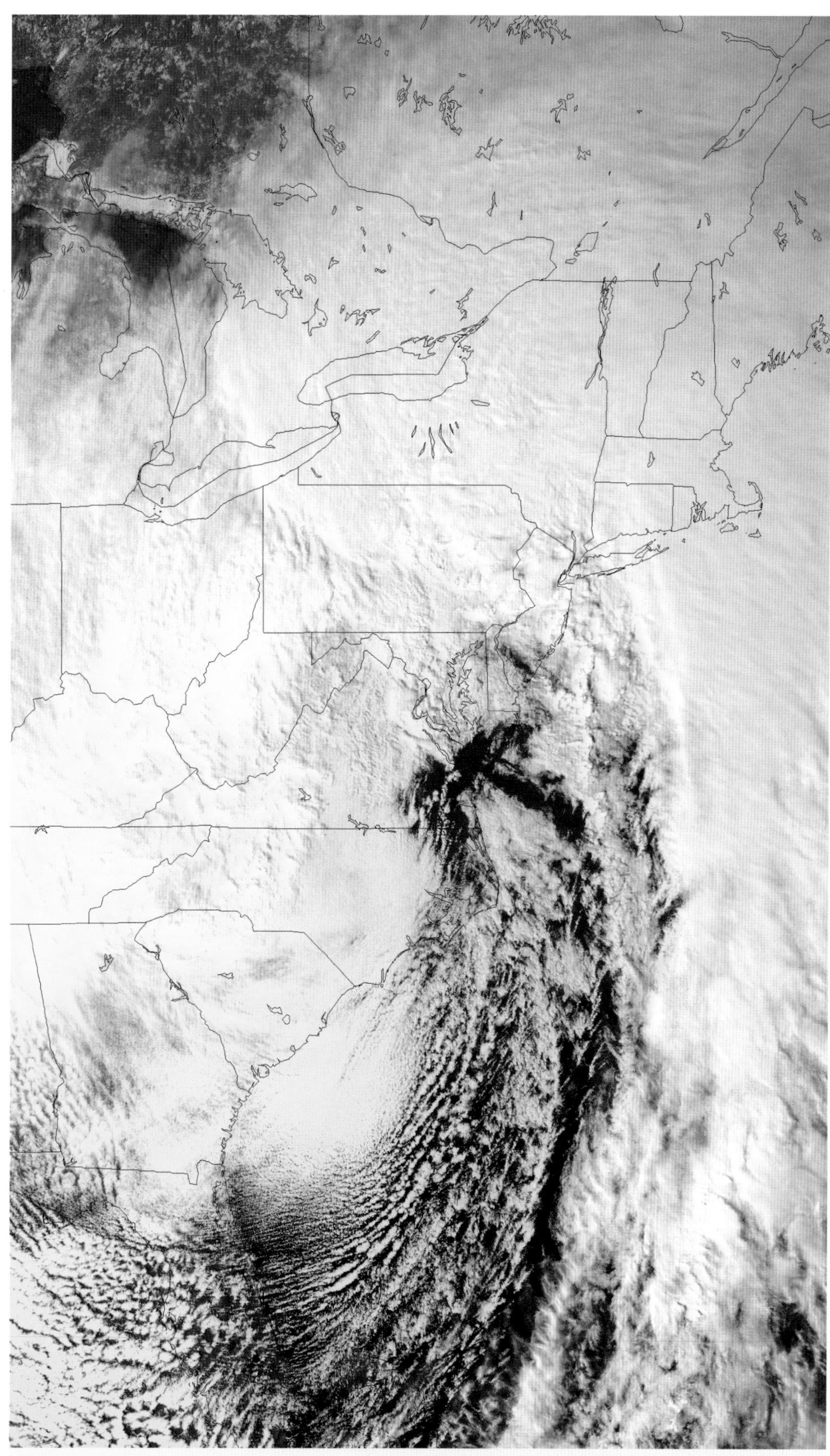

Above: Satellite image of the "superstorm" that engulfed the east coast of the USA in 1993.

Above: Blizzard conditions in the New York streets during the superstorm.

by the mid-evening. It swept in to the Florida coast overnight, bringing with it a succession of eleven tornadoes and killing seven people. Lightning was striking at a rate of 5,100 flashes per hour—during the seventy-two hours of the storm's passage across America over 59,000 strikes were recorded.

On the morning of the 13th, when Jeff Smock and Bill Simmons set out to work on the barn, the newspapers were already warning of a severe storm to come and emergency services throughout the north-east were being put on alert. At the weather observatory being buffeted atop Mount Washington the scene was set for recording the highest surface-wind speed of the storm. High above, winds in the jet stream, turning north-east over Ohio, accelerated to over 200 miles (300 kilometers) per hour, while the center of the cyclone shifted away from Florida and moved north-eastwards up the eastern coast of America.

For 2½ miles (4 kilometers) after they started walking Jeff and Bill were fine; they had been warmed by a short stop at a bar before they had begun their drive home, and when they began to walk they found the going tough, but nothing to be worried about. They were used to the New England winters, and they trudged through little more than 8 inches (20 centimeters) of snow, with just the occasional drift to negotiate. However, Jeff's house was along a track up a steep hill, and as they came over the top they found themselves facing conditions for which they were totally unprepared. "It's hard to explain," said Jeff long afterwards. "We knew right where we were, everything that was going on. Then we got to a certain point, and walking round a corner it went from a foot of snow, all of a sudden, to six foot of snow. The wind blew us down, and our eyelids froze to our face and we couldn't see. You had to hold your hands up so that it wouldn't sting you, the way it was coming down in the wind. It was like the next step was right into the heart of the eye of the blizzard." They had stepped into the peak of what is officially America's worst storm of the twentieth century.

With pressures dropping to a record 960 millibars, the cyclone headed up towards New England and temperatures fell to 14°F (–10°C). Jeff and Bill were in more than 3 feet (1 meter) of snow, wearing jeans, T-shirts, sweatshirts, cotton coats, caps and workgloves; they were surrounded by a howling wind, with snow bearing upon them in a blizzard, bringing visibility down to a few feet, if that, and rapidly disorientating them. They could not go on, and they had nowhere to turn back to. They realized that they needed to find shelter fast, so when they made out a shadow in the white-out they headed towards whatever it would offer. The two large oak trees they had found were to be their home for the night.

Far overhead the jet stream began to retreat back north towards Canada. The storm was at its height and then began slowly to abate, but it would be a full day before conditions on the ground lessened in any meaningful sense. The two men had lost their hats and gloves. Jeff bunched his shirt sleeves round his fists, but in the process exposed his back directly to the elements. Throughout the night they tried and tried to head out for the house, which they knew was so close, but each time they lost their sense of direction or were simply overcome by the snow, and retreated to the trees. The drifts reached 10 feet (3 meters) in height and they dug a half-hearted snowhole for shelter—they even tried to light a fire with matches and hundred-dollar bills. They were exhausted. By morning Bill was semi-conscious and Jeff tried once more to reach help, heading towards a cabin he could just see in the morning light, and which he thought was about ¼ mile (400 meters) away. Bill tried to follow: "I couldn't even walk. I made it about 10 feet from the tree and my feet just gave out. I tried to stand up but I couldn't. I had to crawl back here." His feet were frozen.

Jeff made it to the cabin at noon, and recalls literally breaking ice off the frozen skin of his back while the astonished householders set about warming him with blankets. But Bill was not so lucky; he had another seven hours to wait in the snowhole. During that Sunday, as the low-pressure system gradually began to fill again, the center of the storm passed west of New York City and turned right, directly over Jeff and Bill, to sweep across the state, on towards Maine and out to the Atlantic. Throughout New York State, blizzard conditions remained and Bill lay in temperatures that fell even further, to 7°F (–14°C). When the rescue teams found him, they thought he was dead: he had been exposed for almost twenty-four hours. The two men's injuries were terrible—Jeff had suffered third-degree frostbite burns to 15 per cent of his back and both of Bill's feet had to be amputated—but both recognize that they are lucky to be alive.

Floods, gales and tornados all contributed to the death toll of Superstorm 93, and hundreds of people had to be rescued by the emergency services. The snowfall broke all records. A hundred million Americans, half the nation's population, experienced snow from the storm as more than 1,900 billion cubic feet (54 billion cubic meters) of it fell during the three days of the worst weather in American history.

smog

The cold air of the Arctic does not confine its effects to the great land masses that lie within it and on its border. Cold-air high-pressure systems also move south across the United Kingdom and mainland Europe, bringing the icy cold of the winter north wind. In the 1950s, London was renowned for the "smogs" that befell it in winter, and at the heart of these was cold. The circumstances that led to thousands of deaths from the noxious smogs are perfectly exemplified by the events of December 1952. A high-pressure system moved south over Britain, bringing dry, bitter-cold air and gentle winds. It had been cold and snowing across southern England for weeks. Early that month the winds ceased and the Thames basin experienced a severe temperature inversion, with cold air trapped near the ground beneath a layer of warmer, more moist air above. With the cold air unable to rise through the warmer air above, dense fog formed. At home the people of London began to stoke up the coal fires in their grates, and burned huge amounts of fuel in order to fight the freezing weather outside. This was added to by the many manufacturing industries that still operated within and around the capital, and tons of polluting particles became trapped by the inversion, turning the air into a suffocating cloud of acid rich gas. It's been calculated that each day London produced 1,000 tonnes of smoke particles, 2,000 tonnes of carbon dioxide, 140 tonnes of hydrochloric acid and 14 tonnes of fluorine compounds. And 370 tonnes of sulfur dioxide were converted into 800 tonnes of sulfuric acid as they reacted with water in the atmosphere.

In the sky above, water droplets condensed on the particles of impurities, and the white-colored fog gradually became yellow, amber, brown and eventually black—the "smog"—with visibility falling to just a few feet. Cars were abandoned in the streets, pedestrians were unable to see their way home and in some cases could not even glimpse their feet as they walked. Those struggling through this "pea-souper" had itching skin, their eyes were stinging and their lungs had difficulty drawing breath. But home was not safe either. The polluting air seeped in through doors and windows. Estimates at the time put the death toll from this, the "great smog," at over 4,000, with some 8,000 more dying later of complications. Fortunately today pea-soupers have become a thing of the past, thanks partly to pollution legislation but also to slum clearance, urban renewal and the widespread use of modern central heating.

Opposite: A London Transport inspector uses a lighted taper to lead a bus in a pea-soup fog of the 1950s.
Above: Smog still forms today in some major cities, as here in downtown Los Angeles.

COLD HEALTH

Whether or not it is due to global warming, there is no doubt that, for many people in lowland Britain, snow has almost become a memory and the icy reach of the Arctic cold air is neither so long nor so strong as it was fifty years ago. The harsh winters of 1946–47 and 1962–63 brought parts of Britain to a standstill, and in some areas of the country communities stayed isolated for weeks. Today, as a result of becoming accustomed to warmer winters, the British are perhaps the worst nation in the world at preparing for cold. When weather reports warn of possible snow and ice hospital emergency rooms prepare for their busiest time of the year. But they don't expect that the influx of patients will just be the result of road accidents in icy conditions. Across Britain a 22°F (10°C) decrease in temperature will lead to a 13 per cent increase in the rate of heart attacks and other medical conditions.

In London more people die of cold than anywhere else in Europe. There are 3,129 cold-related deaths per million, compared to 2,457 per million in chilly northern Finland, its nearest rival, and this does not include deaths from winter viruses such as the common cold or influenza. Why is London such an easy victim for cold weather? The answer is alarmingly simple: the British do not prepare for this deadly force of nature and they allow themselves to get cold. In generally colder areas like Scandinavia very simple precautions against the cold, such as wearing gloves and hats, are taken for granted. In London, just walking round Piccadilly Circus on a chilly January day reveals that wearing proper winter clothing is not widespread. And yet such simple measures can make the difference between life and death.

The optimum temperature for chemical reactions to occur in the body is 98.4°F (37°C) and that is why it is your core body temperature. Above it, many enzymes that the body produces break down and cease to function. Below it, the chemical reactions become ever slower, until they stop. The vascular system—the network of veins, arteries and other blood vessels in the body—has an extraordinary ability to control the rate of blood flow around the body's surface and thus control heat loss. Normally blood flows around the surface of the skin at around 300–500 milliliters per minute, but can increase dramatically to 3,000 milliliters to cool you down, or constrict to just 30 milliliters per minute to prevent heat being lost from blood flowing to the periphery of your body. Reaching any body temperature below 98.4°F (37°C) can be regarded as setting in motion the process of hypothermia. At first people begin to stumble, fumble, mumble and grumble, revealing the beginnings of tell-tale changes in motor coordination and mental faculties as the brain and body's chemical reactions begin to slow. By 95°F (35°C), shivering is prevalent, which will generate heat from the chemical reactions involved in the rapid muscular activity. In the short term this can increase heat production by 500 per cent, but eventually the glucose in the muscles becomes used up and total fatigue sets in. By 93.2°F (34°C) speech is slurred and fine motor coordination has gone because the blood flow to the fingers is dramatically reduced—zipping up a coat becomes an impossible task. Shivering becomes more violent; and there is irrational behavior, like trying to take off warm clothing. Below 91.4°F (33°C) shivering occurs in convulsive waves as the body tries to warm itself up, but the energy in the muscles is almost used up and finally the shivering stops. Limbs become rigid as carbon dioxide builds up in the muscles due to the reduced blood flow; the pulse rate slows. By 89.6°F (32°C) severe hypothermia has set in and the body attempts to hibernate, shutting down blood flow, breathing and heart rate; by 86°F (30°C) awareness and consciousness go and the heart may begin to fibrillate; by 78.8°F (26°C) there will be respiratory failure and cardiac arrest.

These are extreme events, but even an unexpected bout of cold weather can be, and often is, fatal. The group most at risk is generally the over-fifties, who may be more susceptible to high blood pressure and run the risk of heart attacks and stroke. They are the unsuspecting victims when the cold strikes. A typical scenario is this. A middle-aged commuter waiting for a bus or train didn't hear the morning's weather forecast. As usual he's wearing his business suit, with perhaps a raincoat for protection. With the "cold snap" that's hitting Britain the weather is feeling distinctly nippy and temperatures are hovering just above zero. As he shivers, the commuter doesn't realize that his body is now going to war against the cold. The blood vessels in his fingers and at the surface of any exposed skin are starting to constrict. Red blood cells are multiplying and as a result his blood is thickening, although he's completely unaware of this. Instead, all he knows is that he needs to go to the toilet more

Above: Winter cold claims many victims because of the failure to take simple precautions—like wearing a hat and gloves!

urgently. In fact, that is a sign of the cold at work on his bloodstream. As the blood vessels constrict and his blood is squeezed, the pressure receptors interpret it as an increase in bodily fluid volume, which normally stimulates the production of more urine. But another effect of constricting blood vessels is a rise in blood pressure. The train is late; he waits for twenty minutes in the cold before it finally arrives; he fumbles for the door handle, which seems very hard to turn as he climbs into the carriage. By the time he gets home he's feeling a bit unwell. After dinner he decides to go to bed early. His body is already showing the symptoms of a heart attack. At one in the morning he wakes, feeling pains in his chest. He's experiencing angina, the sharp pain that results from blood clotting in his arteries. Without immediate medical attention, the man may die.

There are thousands of deaths like this in Britain every year. The awful truth is that they could be avoided by simply wearing a warm hat and gloves.

avalanche

Cold weather, when it does come, can also mean fun for most of us. We love snow and the specialness that it brings—a sense of wonder at nature—and an entire industry is built around pleasure and fun on snow and ice. But ski resorts around the world are always in fear of cold's most awe-inspiring weapon: the avalanche. Born when a mass of snow that has built up on a slope finds that it can no longer stick to the ground, an avalanche can occur with no warning, quietly but with deadly speed, and can shift hundreds of thousands of tonnes of snow in just a few seconds. To be caught in one is a terrifying experience. As the ground is pulled from beneath your feet the air becomes filled with flying ice pounding you, slabs striking you from all directions. Then you are carried along, tumbling as if in a giant washing machine filled with snow—snow up your sleeves, down your neck, inside your clothes, inside every possible entry point to your body; your mouth filling with snow and air as you try to breathe, setting solidly down your throat. Within seconds you could be traveling at hundreds of miles an hour and, within seconds more, brought to a complete halt, fixed within icy concrete.

The worst avalanche on record killed 20,000 people in Yungay, Peru, in 1970. While the scale of that one numbs the mind, the regular falls that occur in the Alps each year are no less frightening. There the weather services provide regular avalanche predictions throughout the winter, in addition to the other kinds of forecast that the public rely on. In February 1999 the services were working overtime—a remarkable set of weather conditions had produced a cluster of avalanches the like of which no one in the Avalanche Prediction Service had experienced. In that month sixteen "level 5" predictions were issued—the most critical conditions—whereas in all the years of its operation the service can recall only three such warnings being issued before. Yet despite their alertness, the staff could not prevent the tragedy that occurred at Galtür, in Austria, on February 23, 1999. In just fifty seconds a wall of snow and ice fell down the familiar avalanche track that the local people knew as the *Wasser-Leiter* (water ladder).

But instead of stopping when it hit the valley floor, it continued on for a further 660 feet (200 meters) to reach the "safe zone" of the village, into which it flowed, filling streets, destroying houses and killing thirty-one people—the worst Alpine avalanche for thirty years. Luggi Salner, who lives in Galtür, recalls the moment the snow struck the village. "I saw a huge wave coming towards us; it was like one of those films about Hawaii. I thought, 'This can't be happening, I'm in the wrong film.' Then I screamed 'It's coming!'" He survived. Christa Kapellner was not so lucky. She and her husband Helmut were returning to their hotel in Galtür: "On the way we took a video; it shows how much snow there was. If we hadn't stopped to take it then, we might have been safe." Instead, the video is a poignant record of Helmut's last moments. "I haven't seen such snow in twenty years," said Christa. 'The avalanche came from behind. Helmut shouted 'Christa,' but I couldn't answer. His legs were pulled from under him by the force, and I was hit from behind." Christa was found by a rescue dog and brought out alive. She had been carried round a bend in the road and was buried under two cars that were piled on top of each other. Helmut had been carried a further 165 feet (50 meters) by the rush of the snow, and was dead.

The avalanche had been caused by one of those freak combinations of weather conditions that conspire to bring what we least expect. The prevailing weather pattern that flows over the Alps is driven by the Atlantic Ocean. It is

Above: An avalanche in mid-flow down a Peruvian mountain valley.

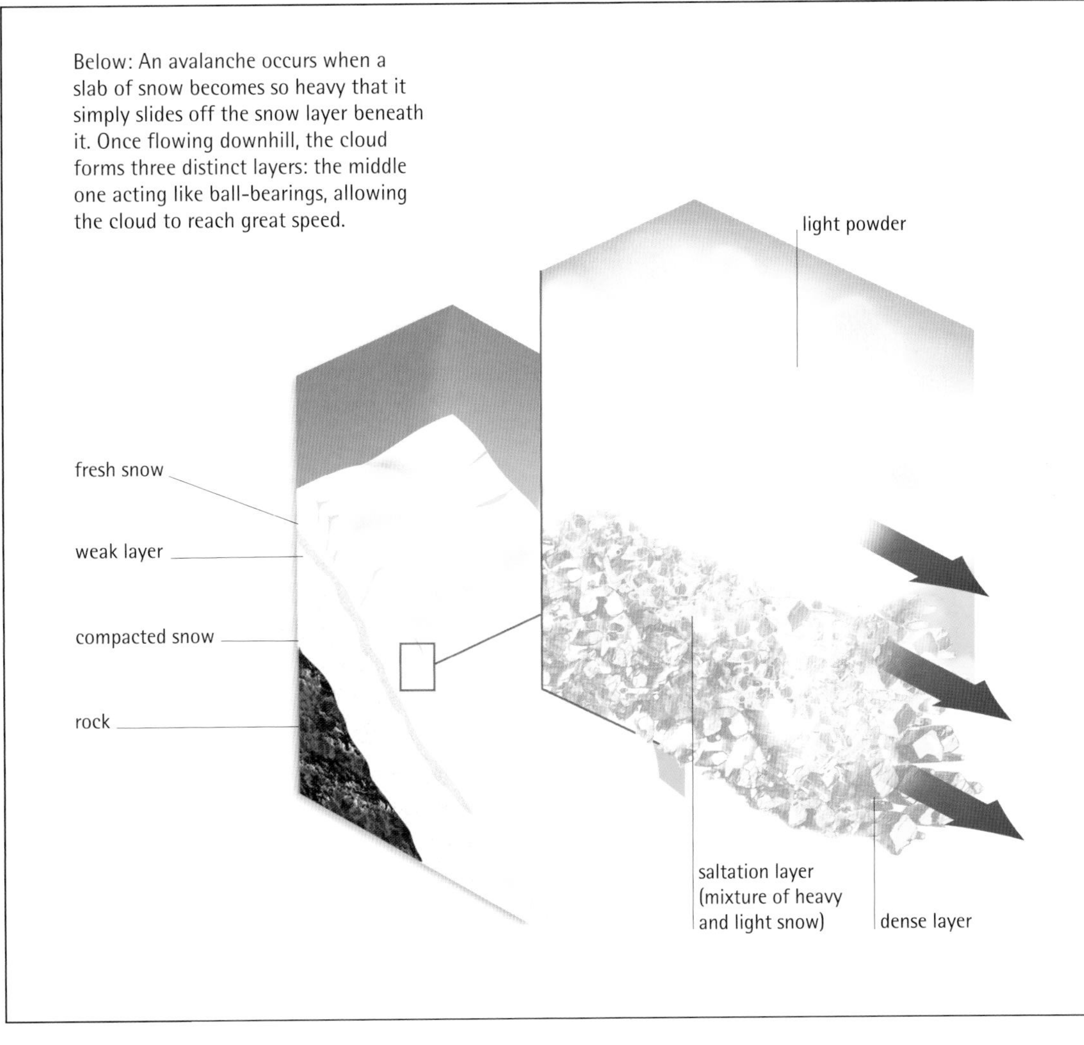

Below: An avalanche occurs when a slab of snow becomes so heavy that it simply slides off the snow layer beneath it. Once flowing downhill, the cloud forms three distinct layers: the middle one acting like ball-bearings, allowing the cloud to reach great speed.

known as the "weather kitchen" for the Alps, because it provides very moist air that comes off the sea, flows in across France and rises as it meets the mountains. Here it cools and creates the plentiful snow that the ski resorts need. In February 1999, however, a high-pressure system developed off western Europe and the Atlantic winds were deflected to the north.

A storm had developed from a mass of warm air that had flowed up from the tropics, and this turbulent mass of moist air was driven north with the wind. There it wheeled round to join the freezing Arctic air before turning south to flow down the North Sea, now carrying an icy blast across the Low Countries before striking the Alps broadside. At this point the full width of the mountain range presented a block to further progress across the entire weather system. The result was that the storm remained over the mountains for days, depositing vast quantities of precipitation. And then a succession of storms followed, each taking the same path, each lingering at the mountains and each relentlessly creating snowfall. Parts of Switzerland and Austria experienced six times their normal amounts of snow. Throughout January it fell, building up to an enormous depth on the slopes above Galtür. Then strong winds, up to 60 miles (100 kilometers) an hour, whipped up more of the fallen snow and redeposited it on top of other drifts. Wind is known as the builder of avalanches, and at Galtür the north-easterlies blew for three solid weeks in February, piling up the snow on the slopes directly above the village.

The key to whether a vast amount of snow eventually slides downhill as an avalanche or simply rests harmlessly on the slope is the nature of the snow crystals themselves. Angular-shaped crystals will interlock and bind well together; pellets or granules will slip and fall. What's known as a "slab" avalanche occurs when one layer of snow, perhaps fresh and well tied together, slides away from a layer beneath, which is typically older, more compacted snow with a smoother surface. The layer between them, at which the sliding begins, is known as the "weak layer" and normally it gives way long before the weight of snow reaches the proportions found at Galtür. So smaller avalanches occur sooner, preventing a major one from developing at all. But at Galtür the weak layer stayed in place for an unusually long time, allowing much more snow to build up above it, so when the slide finally occurred the scale of the avalanche was all the more terrible. When investigators began to unpick the events that led to the disaster, they could not understand at first why the weak layer had held on for so long, but when they analyzed this part of the snow they found frozen meltwater between the grains. This "melt crust" represented a layer that had thawed out at some stage during the day only to refreeze at night, which meant that the crystals of the layer bonded together more strongly, making it all the more long-lasting. When the meteorologists looked back at the weather records, they saw that at the end of January a brief warm spell had occurred between the two main waves of snowstorms. In this tiny weather window the melt crust had formed—a full month before the disaster—strengthening the slope, allowing the snow pack to build and sealing the fate of the village below.

Scientists at the Swiss Snow and Avalanche Institute have recently developed an extraordinary experimental technique to try to understand what goes on inside a flowing avalanche. They have built a concrete bunker and packed it with instruments so that they can detonate a slope of snow and allow an avalanche to pass right over them while they watch. This was first tried only two weeks before the Galtür event. For some time scientists have known that there were two layers to an avalanche: a huge cloud of light powder snow swirling above a denser layer below. But the bunker experiment has confirmed the existence of a third layer—known as the saltation layer—a tumbling mixture of light and heavy snow that flows in the middle. It acts very much like a layer of ball bearings, increasing the speed with which the one layer can flow over the other and allowing large chunks of ice to roll over smaller ones at speed, to devastating effect.

When it gave way the starting zone of the avalanche at Galtür was about 1,600 feet (500 meters) across and, at first, snow to a depth of about 10 feet (3 meters) began to slide. About 170,000 tonnes of snow broke loose, but in the fifty seconds this took to come down the mountain it doubled in size, eating into the layers of untouched snow below it and on the sides of the valley, so when it reached the valley floor a third of a million tonnes of snow and ice, traveling at 200 miles (300 kilometers) an hour, were a few seconds away from the village. When it hit, in a wall 300 feet (100 meters)

Right: A cloud of powder snow at the top of an avalanche on Mount McKinley, Alaska.

high, the icy chunks of the saltation layer did the damage, while people were suffocated by the powder snow that rode above it.

It was an exceptional set of climatic circumstances that produced the Galtür disaster. The little village should have been safe, but once the weather pattern had set in nothing could have prevented the tragic outcome.

meltdown

Of all the elements, cold remains our deadliest enemy. Our bodies evolved in the heat of Africa and our civilization has thrived in a period when cold has been in retreat, with the glaciers of the last ice age slinking back towards the poles. But each year we survive its advance, and each year the orbit of the Earth around the sun allows the seasonal cycle of winter and summer to take its course. Each year the threat of the ice and snow recedes as the warmth of summer begins to set new challenges, challenges that come with the other great force of the weather—heat. In the next chapter we will see that humans also have a limit at the other end of the temperature scale; and in revealing the weather that comes with heat we come face to face with the ultimate energy source that keeps us alive—the sun.

Above: The ruins left behind by the Galtür avalanche, the worst such disaster in Austria for over 40 years.

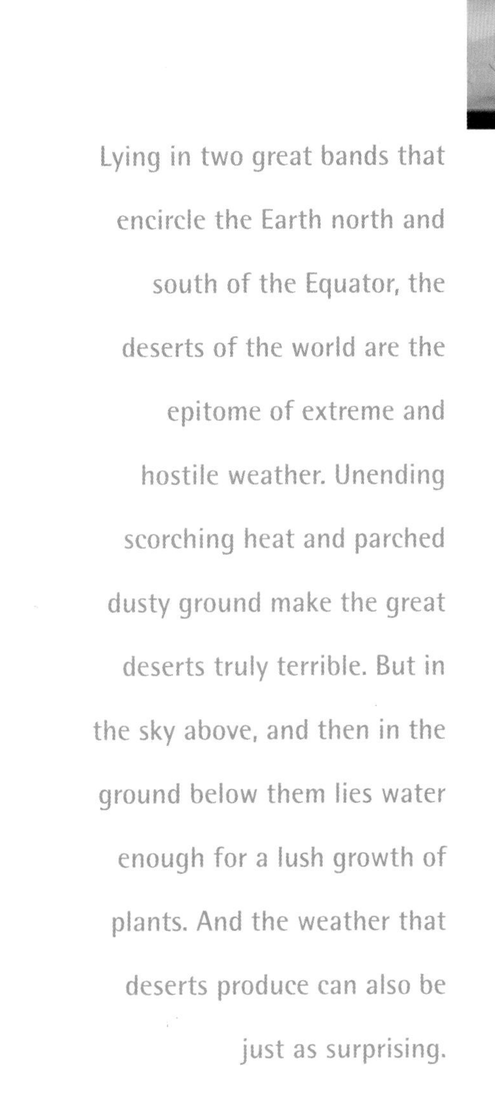

Lying in two great bands that encircle the Earth north and south of the Equator, the deserts of the world are the epitome of extreme and hostile weather. Unending scorching heat and parched dusty ground make the great deserts truly terrible. But in the sky above, and then in the ground below them lies water enough for a lush growth of plants. And the weather that deserts produce can also be just as surprising.

chapter five

hot

design for the desert

One living creature has become so well-adapted to the heat that at the rear of its head there are special blood vessels designed so that the circulation keeps the brain relatively cool, even when the rest of the body has reached temperatures of a crippling 113°F (45°C). A special store of body fat can provide nourishment for up to six months, and the cells can tolerate a level of 1.8 per cent salt in water, a concentration that would prove fatal for a human. When it drinks, the creature can drink 26 gallons (120 liters) of water in ten minutes. And whereas a man will lose 5¼ pints (3 liters) of fluid an hour walking in extreme heat, perhaps shedding 10 per cent of his body weight in a single day—which would lead rapidly to death—this animal hardly sweats at all. But, as liquid is eventually lost, and it does become dehydrated, it adjusts the volume of its vascular system to reduce the work that its heart must do; and the filtering action of its kidneys can be cut by 20 per cent so that it loses only a thousandth of its body weight through urination each day. Its ultimate trick is that it has an extraordinary ability to combine hydrogen with the oxygen from the air that it breathes, to make its very own water. Glands also supply its eyes with extra fluid to keep them moist, while long lashes keep them shaded and the eyelids are so thin that it can see where it is going even while they are shut, bringing protection against flying dust. Its feet are ridiculously broad with padded toes that prevent it from sinking into sand; its nostrils can close completely or leave just a tiny hole for breathing; and its ears are far to the back of its head, and covered with protective hair even on the inside. It can hear well, but it is famed for paying little attention to commands.

It is, of course, a camel, and it has evolved to live in the most extreme of conditions. It is the very epitome of the desert.

Above: Tiznit, an oasis town on the edge of the Sahara in Morocco.
Opposite: Contrasting landscapes in the Mojave Desert, California.

A man walking in extreme heat will lose perhaps 3 liters or 5¼ pints of fluid every hour ●

Above: The character of a desert is dictated by the nature of the underlying rock. Here, wind erosion has left white sands in Mexico.

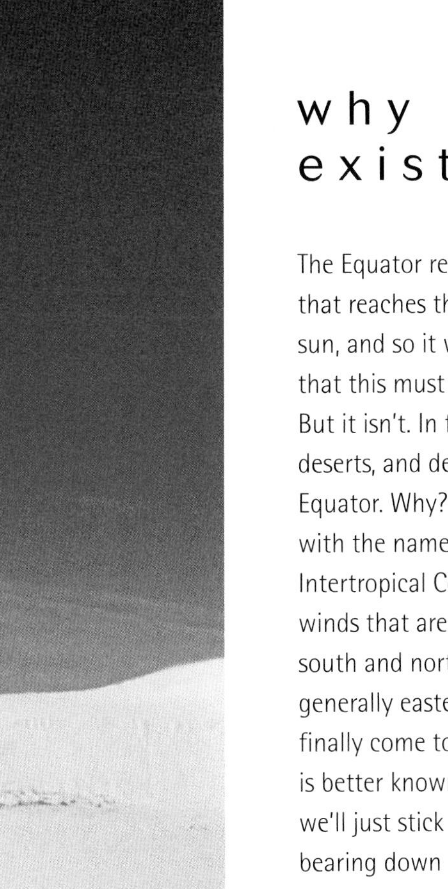

why deserts exist

The Equator receives the lion's share of the heat that reaches the surface of the Earth from the sun, and so it would be common sense to assume that this must be the hottest place on the planet. But it isn't. In fact, the hottest places are the deserts, and deserts are not what you find on the Equator. Why? The key lies in that magical region with the name that rolls off the tongue—the Intertropical Convergence Zone (ITCZ), where the winds that are flowing towards the Equator from south and north, and in both cases from a generally easterly direction (see Chapter Two), finally come together in a zone of calm. At sea it is better known as the doldrums, but on land we'll just stick to ITCZ. The heat of the sun bearing down on the equatorial surface is very great and the air that converges there has to move somewhere, so it rises as it warms, leaving an area of low pressure behind to take in more air from the converging winds. Warm air can hold much more moisture than cold air, and, as it rises to a very great height, it cools. The moisture in the air begins to condense to cloud droplets and towering rain clouds are formed.

Out over the oceans these clouds begin to rotate and spiral together, giving birth eventually to the tropical cyclones and typhoons that track the tropical seas. But over the great continents of South America and Africa that straddle the Equator, the clouds do not spiral up into hurricanes. There simply isn't enough water to fuel them. Instead, they settle for rain: lots and lots of tropical rain. And the rain is eagerly soaked up by the roots of every piece of growing vegetation, which already has a head start because of the intense heat that is bearing down on it

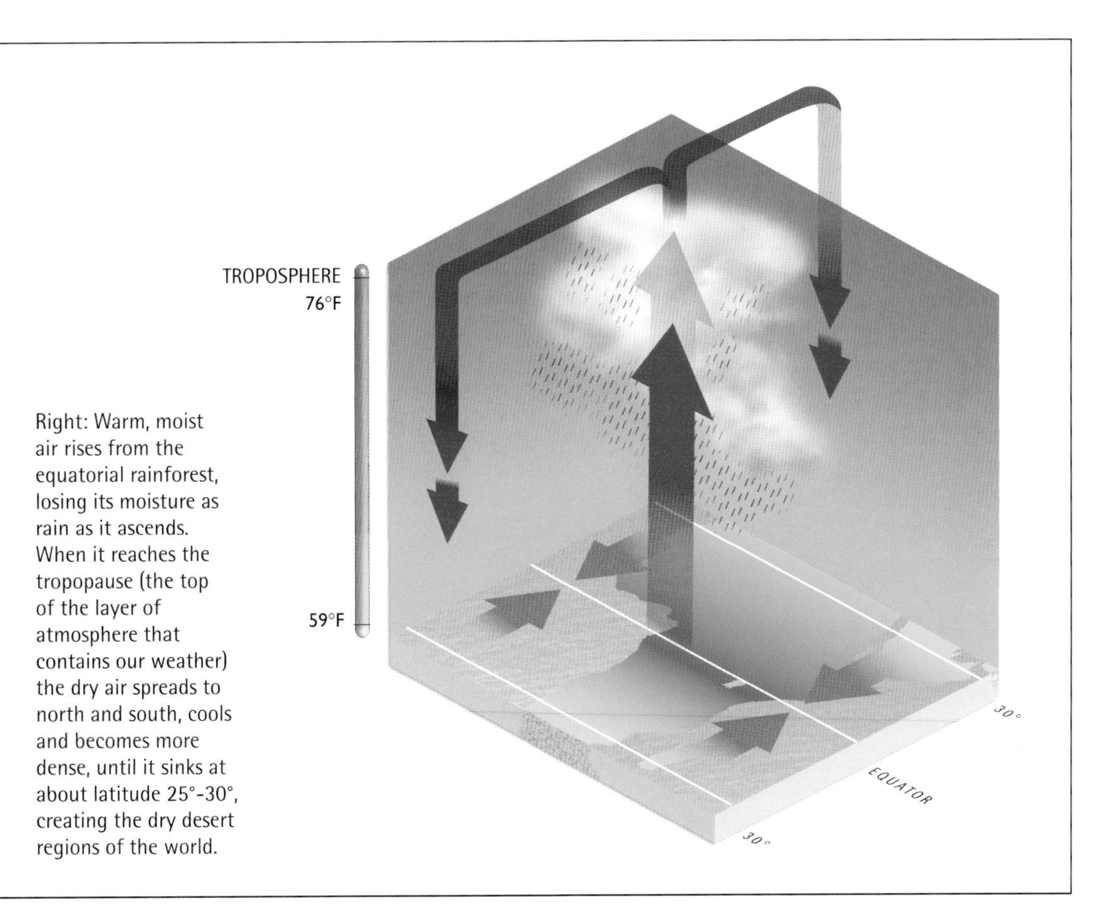

Right: Warm, moist air rises from the equatorial rainforest, losing its moisture as rain as it ascends. When it reaches the tropopause (the top of the layer of atmosphere that contains our weather) the dry air spreads to north and south, cools and becomes more dense, until it sinks at about latitude 25°-30°, creating the dry desert regions of the world.

throughout the year. It is growing season all year round at the Equator. The plants themselves then fuel the air with more moisture as water vapor is transpired out of countless leaves, ferns, fronds and every possible variety of greenery, to add to the extraordinary humidity of the region. To stand still in an equatorial jungle is to perspire without expending a single additional morsel of energy.

So rain and cloud and rising warm air are locked in a virtual feedback loop that creates the rich variety and lushness of the world's tropical rainforests. But as the warm air climbs to almost incredible heights, shedding deluges of water from billowing clouds, it eventually reaches the top of the troposphere—the layer known as the tropopause—where it begins to move away from the Equator towards the poles. This is the beginning of the great global circulation of air, described in Chapter Two, which is the engine of the weather. Driven on by the endless cycle of air rising below it, the high-level air has now been stripped of its water. As it flows north and south away from the Equator, it cools, becoming more and more dense until, at a latitude of around 25–30 degrees north or south, it has become heavy enough to fall as a flow of dry air towards the Earth. As it falls it becomes more compressed, and begins to warm up, at a rate of about 22°F (10°C) per 3,000 feet (1,000 meters) of descent. The warm, dry air builds high pressure, with little moisture content to create clouds. This region is the subtropical high-pressure zone, and it is what sets the conditions for a desert.

In fact you only have to look at an atlas of the world and it all becomes startlingly obvious. Laid out before you are two clear bands a little way north and south of the Equator. Along these almost all of the world's deserts are spread out, marking where the warm, dry air descends and begins to make its way back either towards the poles or towards the ITCZ, to help drive the cycle all over again.

Left: Clouds of moisture hang in the canopy of the hot Amazon rainforest.

rivers in the sky

As the sun's energy bears on the tropical region of the planet, so it warms the surfaces of both land and sea. Over the oceans the water warms and evaporates and vapor rises, producing the warm, moist air that will give birth to the tropical storms of the region. But something else happens that is quite remarkable, and has been discovered only within the last decade. As the water vapor rises steadily it encounters various different low-pressure disturbances and frontal systems emerging from the general wind patterns in the tropics. In certain parts of the globe these conspire to collect the water together to form huge streams of invisible water-vapor—rivers in the sky, if you like—that begin to flow across the world. Higher than 13,000–16,500 feet (4,000–5,000 meters) above the surface of the sea, the air becomes too cold to hold much water vapor, so the rivers in the sky form below that altitude. These atmospheric rivers are vast, perhaps 250 miles (400 kilometers) across and stretching for 3,100 miles (5,000 kilometers) in length. Typically, these transport as much water as the River Amazon—the difference being that they are in the sky. Climate scientists first discovered them when using satellite data to try to track the movement of the "greenhouse" gas, carbon monoxide, from forest fires in Africa across the southern Indian Ocean to Australia. Instead they detected high concentrations of water vapor in five major streams in the southern hemisphere and five in the north. The flow of these streams across the planet shifts and changes with the seasons, but typical journeys are: from Australia and Indonesia, north across the Equator and west to the Arabian Sea, turning towards India in the monsoon season; from southern Japan across the Pacific to Seattle; from Puerto Rico in the Caribbean up over the Atlantic to Iceland, Norway and Greenland; and from Florida across to Ireland, or turning to Spain in the summer.

As these low-level rivers of vapor are invisible, it is impossible to see the actual path of their flow directly, but satellite sensors have picked out their route and photographs from space have also hinted at their paths, by identifying cloud formations where water vapor rising above them has begun to condense into clouds. It is clear that the rivers in the sky can penetrate storms, cyclones and hurricanes, becoming drawn into the storm system, condensing out some of their water and intensifying the weather that they have collided with. However, the study of these rivers is still too new for us fully to understand the role they play in the weather system as a whole.

a desert world

A view of the Earth from space reveals the extraordinary extent of the world's deserts. They lie spread out in stark contrast to the green areas to their north and south. In the southern hemisphere, along the imaginary line called the Tropic of Capricorn, lie Namibia's Kalahari, the Gibson and Simpson deserts of Australia and the Monte desert of Argentina. While, across the Equator to the north, lying along the Tropic of Cancer, are the deserts of India, Iran, the Arabian peninsula and, of course, in North Africa, the mighty Sahara. But there are other deserts, which can form in different ways; the "rainshadow" deserts, such as the Mojave or most of the states of Nevada and Utah in the USA, or the Patagonian desert in Argentina, are created because they are on the leeward side of high mountain ranges, which rarely allow any moisture from the prevailing winds on the other side to cross them. There are also continental interior deserts like the Taklamakan in China or the vast Gobi that spreads across Mongolia, which is so far from a coastline in any direction that it is hard for moisture to find its way there. And there are others that are tantalizingly close to the ocean but uniquely starved of water. One of these is along the shoreline of Peru, where the ocean currents in the sea cause a great upwelling of cold water next to the coast. This cools the air above it which cannot then hold much moisture, and what little there is rolls in as banks of coastal fog. It never rises high enough to become rain clouds, so the land is deprived of rain. The Atacama Desert is the result. Incidentally it is this same ocean current that, when disturbed, gives rise to the dramatic climatic event of an El Niño (see Chapter Three).

No desert is truly dry. Deep underground there is always flowing water, which has soaked into the ground from rainfall perhaps thousands of years before, and thousands of miles away from where it finally reappears. The water travels through a layer of porous rock, an aquifer, drawn essentially downhill by the action of gravity, until it reaches a point at which it can go no further, perhaps due to a band of impervious rock along a fault line. Then the huge pressure behind it causes the water to rise to the surface along the fault line, to create an oasis. Other oases appear where erosion has worn away a layer of rock and exposed a porous layer directly to the surface. However they are formed, they are the wonders of the desert, enabling people to survive and eke out an existence among the eroded rocks and wind-blown dust around them. But an oasis is almost the only source of water in a desert, and if you do find yourself in one in the Sahara, it is perhaps worth reflecting that the pure, cool water you drink may have rained on the ground during the time of the pharaohs.

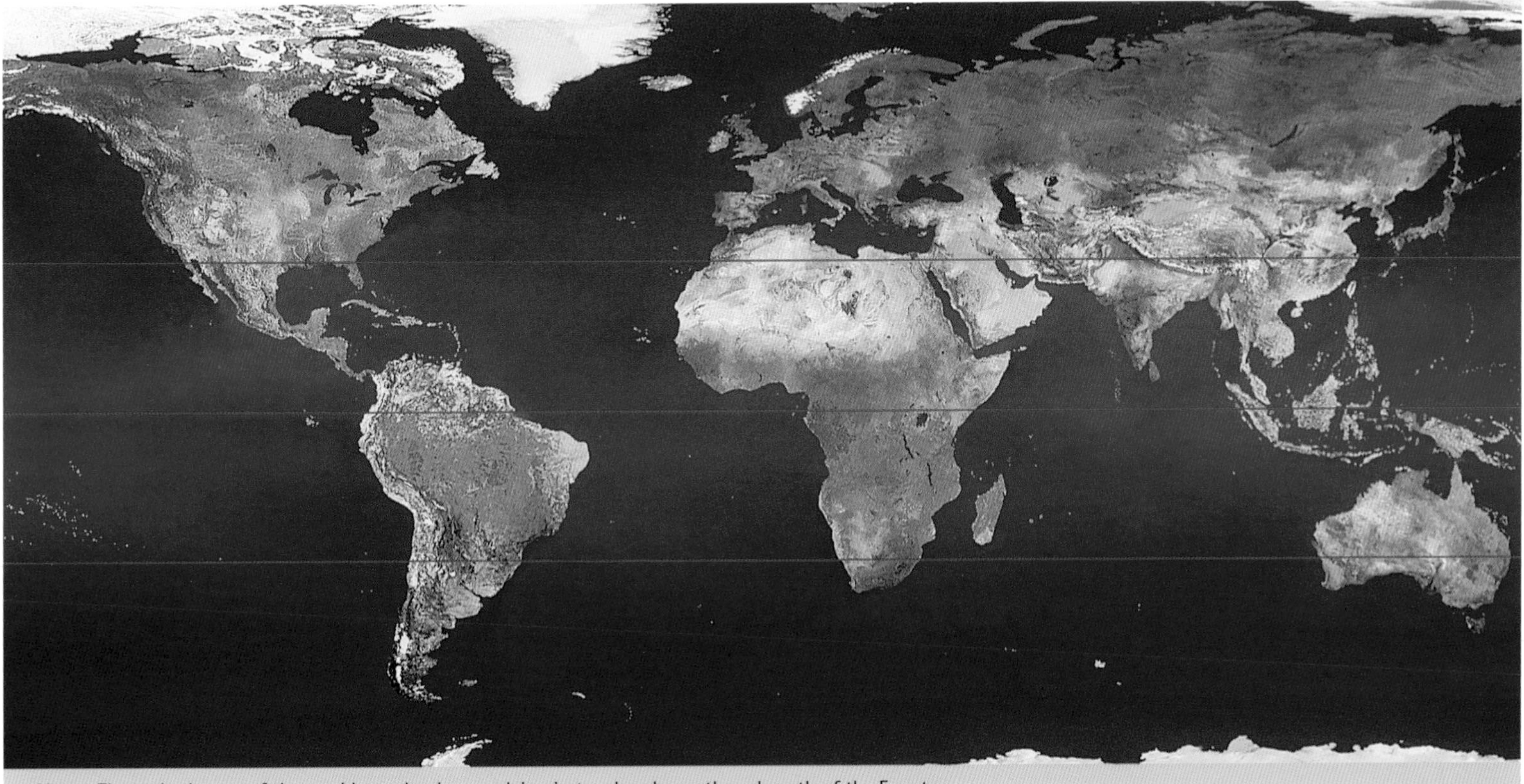

Above: The main deserts of the world are clearly seen, lying in two bands, north and south of the Equator.

THE HIDDEN PAST–SAHARA

Space exploration has produced extraordinary insights into the workings of our planet, and when the space shuttle Columbia had recovered from the first flush of its successful flight in March 1981 and its second mission was underway, the astronauts on board began the kind of experiments in imaging the Earth that have become a feature of low-Earth orbit over the last twenty years. In November 1981 they used a new imaging radar system to look at the western desert of North Africa, part of the Sahara, in order to map its surface and extent accurately. The area they were looking down on was a region of deep sand, known as the Selima sand sheet; but when the images were processed the US Geological Survey scientists who viewed them were in for a surprise. The sand sheet did not appear to be there at all. In its place was a strange terrain of valleys and hills, rivers as wide as the Nile and channels that flowed into them—a very different world. It seemed that the Selima sand was so dry that it was completely transparent to the radar, which depended on at least some moisture in the soil to absorb and reflect the radar signals. The result was a snapshot of the Sahara as it had been, some thirty-five million years ago. Today, the Selima sands perhaps receive no rain in a decade or more. But here was a long-buried array of once-fertile river valleys and lush vegetation; plains that in times gone by had been wet with rain, with

powerful streams carving gullies through the bedrock.

The world's deserts are constantly changing. As recently as the Ice Age the deserts of the mid-latitudes expanded, and large dust clouds blanketed Asia and Europe—with so much moisture trapped in the ice, the climate as a whole was much drier. And within the period of human civilization the deserts have also changed dramatically. Cave paintings in the heart of the Sahara reveal that the people who once lived there hunted elephant, giraffe and antelope, and pollen samples have revealed that oak and cedar forests were widespread. Some six to nine thousand years ago it was the "green Sahara" with lakes and vegetation. This was due to slow changes in the orbit of the Earth around the sun and a general warming of the northern hemisphere. A warmer planet means more moisture evaporating into the atmosphere, and so more rainfall across the lands, and the warming landmass brought a greater intensity of the seasonal monsoons. The Sahara flourished.

But then, five thousand years ago, the change began: over barely a century the monsoon collapsed, although the edges of the desert remained lush enough for the Romans to colonize North Africa, where the northern fringes of the Sahara became the "breadbasket" of their empire.

Left: A view of the Sahara desert in Sudan from space, showing the orange color of the dry earth.
Below: Image from space of the Selima Sands, overlaid with a band of the same image reprocessed, to reveal the ancient rivers and lush valleys that once existed in the Sahara.

Right: A computer-generated image of a huge dust cloud moving over the western Sahara and spreading out over the Atlantic Ocean towards the Caribbean.

killer desert

In the most extreme, arid deserts, such as the eastern Sahara, the rate of evaporation from the ground is two hundred times the rate of what little annual rainfall there is. The skies are clear; it has been calculated that of the 4,100 hours of annual daylight in the center of the Sahara, 3,978 of them are filled with direct sunlight. During the day heat is unrelenting. July temperatures are regularly over 113°F (45°C) and the highest ever recorded, at El Azizia in Libya, was 136.4°F (58°C) in the shade. These are temperatures that will kill a human. The body's normal core temperature is maintained at 98.4°F (37°C); if it rises to 106°F (41°C), heatstroke will have set in; above 108°F (42°C) a human being will die. So our bodies have developed a sophisticated set of responses to keep that all-important temperature steady: it's called sweating. As our three million sweat glands slip into action the evaporation of sweat cools the skin, so the blood circulating near the surface also cools and returns to the core of the body, helping to cool it down. At its maximum human sweating will lose 5 pints (3 liters) of water per hour, and that is when the desert can kill. As the body loses water it draws water from its cells, so the chemical balance within them is disturbed, causing damage to proteins. These are involved in the chemical signaling that goes from cell to cell, so gradually the actions of the body and the messages within the brain begin to go awry.

Lose 5 per cent of your body water and fatigue and dizziness set in; 10 per cent loss will bring delirium; and over 15 per cent is invariably irreversible. The tongue swells, the skin shrinks and cracks against the bones, the eyes sink into the skull. Finally, the blood thickens with lack of water and so ceases to circulate properly. There is a sudden rise of core temperature and death. A few hours in the Sahara can bring on such a result, which makes the extraordinary adaptation of the camel all the more remarkable. Its long legs keep its body away from the baking sand, its hairy coat provides insulation from the blazing sun and its remarkable hump provides a store of fat that will convert to water, allowing it to lose 25 per cent of its body water without any adverse effects.

The people who have adapted their lifestyle to the extremes of the desert have done so in both striking and practical ways. The Bedouin and Tuareg tribes that travel the length of North Africa are made up of tall people who have adapted physically to the heat. Being tall and slim enables the body to maximize the surface area available for sweating, while at the same time minimizing the surface on which the sun beats down directly—the head. These tribespeople live their lives covered from head to toe, to protect against wind-blown sand and dust but also to stay cool. One of the apparent inconsistencies of their attire is that black is a favoured color. Yet black absorbs far more heat than does white. So why do they wear it? The secret lies in the many layers of clothing in which they wrap themselves. The black robes at the surface are 11°F (6°C) warmer than white ones would be, but the layers of clothing beneath provide effective insulation between that surface and the body. And the warmer surface means that air convection inside the clothing layers is more rapid, so that fresh air is drawn in from below and rises up through the layers of air, keeping the body cooler.

desert survival

The proximity of the Great Sandy Desert of Australia to all the trappings of twenty-first century civilized living in the farmsteads and towns at its edge has meant that many people have been lulled into a false sense of security when venturing into its clutches; and some of the most remarkable, terrifying and poignant stories of human survival and death have been played out in its vastness. In 1999 a huge rescue operation led by an American expert tracker and a private rapid-response organization was set under way to try to find Robert Bogucki, a young American who had set off to cycle through Australia with his girlfriend, Janet North. When Janet pulled out because she could not keep up, Robert continued alone, and the last postcard he sent to his parents told them he wanted to cross the desert, and not to expect to hear from him for a few weeks. In late July his bicycle was found at a roadhouse on the main road that runs along the edge of the desert and along Australia's western coast, and a search was organized. After two weeks and no signs, Robert's father called in the private rescue-organization and they set off, accompanied by a posse of journalists, with the

Left: The gypsum dunes of White Sands, in New Mexico, USA.

unspoken mission of finding his son's body. Astonishingly, his trail was picked up over 60 miles (100 kilometers) further north. All the signs were that he was still alive and managing his walking and resting in the most efficient way that he could, sheltering from the sun in holes and surviving on the infrequent saplings and desert flowers. Bogucki was lucky; he had strayed during the Australian winter. He was found in the end, crouched next to a muddy pool by a television news helicopter that had been following the search, nearly 110 lbs (50 kilograms) lighter in weight than when he had set out. He said afterwards that he had gone into the desert deliberately, to "sit down and let my mind go free."

For others, the experience of trying to survive desert conditions was more horrific. Matthew McCough and his young daughter Shannon were on a camping trip to remote western bushland when their car broke down. After several days their spare water had run out and they were reduced to drinking their own urine and car-radiator fluid. In the end Matthew was on the point of acting on a terrible decision. He had realized that he would have to kill his daughter, and then commit suicide, rather than see her suffer or have to die alone if he should perish first. It was a choice that no parent should have to make. They, too, were lucky: a passing gold prospector found them just in time.

James Annetts and his friend Simon Amos were not so lucky. On holiday from their jobs as ranch hands, the two teenagers ventured into the desert in December, the height of the scorching summer, and got lost. Their skeletal remains were found weeks later amid poignant messages to their loved ones that they had scratched on their empty plastic water-bottles. One body had a bullet hole in the head, although whether from suicide or at the hand of a friend's mercy killing will never be known. The boys had used car tools to make a crude SOS sign on the roof of their abandoned jeep which was found 9 miles (15 kilometers) away from their remains, and they had placed sticks on the sand in the form of an arrow to indicate which way they were heading on foot. It appears that, once in the desert, they had begun to drive in circles, become bogged down in sand and drained the vehicle's battery in attempts to get it restarted.

making a desert

When the blazing sun goes down at the end of a desert day, the baking ground begins to radiate its warmth straight back up into the air. With no cloud cover to retain the heat, that process is rapid and dramatic: the temperatures in a desert fluctuate wildly between day and night, from over 122°F (50°C) in the afternoon to below freezing in the small hours of the morning. As the rocks, stones and baked soil of the desert expand and contract in this daily cycle, so they become fragile and fractured and crumble away. With no vegetation to protect the ground from the wind, the process of erosion by blasting from sand and dust is almost continuous. A desert is effectively self-perpetuating. Ironically what little water does fall to the surface only increases the destructive process. It evaporates to leave salts that add to the chemical weathering, or freezes overnight and expands inside tiny fissures in the stones to add to the cracking. In fact, the stresses placed on desert rocks by these temperature extremes can have spectacular results. On one of his journeys through the Sahara in the 1970s, the German explorer Uwe George reported sitting next to a large boulder, about 3 feet (1 meter) wide, when it suddenly exploded with the roar of a cannon, shattering into a thousand pieces.

Although deserts are perhaps famed for their sands, most of them are largely weathered

Above: Weathering of sandstone has left behind dramatic columns and towers in the Tassili Plateau in Algeria.

Above: Wind blowing over dunes in the Nefta Desert in Tunisia.

rock and stone. The Sahara, for example, is only 20 percent sand. However, it has to be admitted that this 20 percent makes up one of the great natural wonders of the world—the sand seas, where slow-moving waves of dunes range across areas the size of whole nations. The sand is the product of the final stages when rock weathers and is created over millennia. The Great Eastern Sand Sea, the Eastern Erg, in Algeria was probably laid down over some ten thousand years and consists of dunes ranging from a few feet to over 1,000 feet (300 meters) in height. It is added to every year. The creation of sand from rock and gravel is known as "liberation." This process can be witnessed in Algeria where the wind that whistles down off the Atlas Mountains scours out the desert floor. It blows the sand across the country until it reaches the basin in which the Eastern Erg formed, adding 6 million tonnes of sand to it each year that passes.

The Sahara is 3,100 miles (5,000 kilometers) east to west and 1,000 miles (1,500 kilometers) north to south, almost 3,500,000 square miles (9 million square kilometers) of suffocating heat and virtually constant wind. It is roughly the same size as the USA and yet, unlike the USA, it is growing. Nothing symbolizes the power of the desert more than the sight of an oasis that has waves of sand slowly burying its trees as a sand sea slowly advances. In fact, the encroaching movement of the world's deserts is far more insidious. Over-grazing, over-farming, over-mining—all these factors of human activity add to the gradual destruction of usable land on desert fringes. A third of the planet's land surface is already desert and that proportion is slowly increasing. Twenty-three thousand square miles (60,000 square kilometers) of land are lost each year—over 58 square miles (150 square kilometers) a day—to the processes of desertification.

SECRETS OF THE DUNES

To climb to the top of a large sand dune is an exhausting process; the sand slides from beneath your feet, burns your skin with the heat reflected off it and tires you with the steepness of its slope. But to look over a sand sea from such a vantage point is a staggering sight. Dunes appear in many different formations and these reflect the supply of the sand of which they are made and the pattern of the winds that blow on them.

TRANSVERSE RIDGES

The simplest dunes form in the lee of a boulder, a shrub or a hill, while others form in hollows as sand is blown into them. With constant wind and plenty of sand, a dune can build up over a very long length at right angles to the wind: one dune in Mauritania is over 60 miles (100 kilometers) long. But because of the way a mound of sand disturbs the air that flows over it, a dune always tends to be replicated. The wind carries a mass of small grains of sand which "jump" through the air. The heavier grains move slowly along the surface of the ground, pushed along by the constant bombardment of the jumping grains. The wind blows quickly across hard, pebbly ground, but when it reaches more sand it slows because of friction, and the grains that are jumping through the air are deposited on the ground, more and more of them, so that a dune builds up. When it reaches a certain height, the ridge becomes exposed to the wind and the small grains start to be blown over the top, tumbling down the other side to start the process again. The wind sweeps the new loose sand up into a new ridge in the lee of the first one, and a sand sea is born.

LINEAR DUNES

Where the winds blow strongly and steadily from one direction a different kind of long dune can form. The sand is cut with long grooves parallel to the direction of the wind and piles up on either side, creating dunes that can run for hundreds of miles in long, parallel stripes across the desert. The Simpson Desert of Australia is home to some of the most striking examples. It is not clear exactly how the evenly spaced stripes are created, but one theory is that the wind is set up like a series of rotating corkscrews as it flows along the ground, with the sharp edge of the dunes forming along the line where each parallel corkscrew meets the next one.

BARCHANS

Where the wind blows from one direction but fluctuates in strength from place to place, a barchan forms. It is a smaller, crescent-shaped dune with its tips pointing downwind. These smaller dunes, constantly battered at their edges by fluctuating wind, with the sand continuously rolling down the leeward slope to pile up against the next ridge, are the most mobile. They can move across the desert at up to 165 feet (50 meters) a year.

Left: A field of crescent-shaped barchan dunes in the southern Sahara.

SEIFS

Seif, the Arab word for "sword," is used to describe a type of dune that has a sharp crest. A seif forms where sand is plentiful and creates short, snaking ridges parallel to the direction of the wind.

STAR DUNES

These beautiful shapes, like giant starfish lying far up a sandy beach, form where the wind direction is very variable. A dune and crest form, and then part blows off at an angle as the wind changes. Effectively, star dunes are like several barchans that have blown together. Because there is no consistent wind direction where they form, star dunes do not move across the desert. Instead they can build to very great heights as the sand blows up against each side in turn.

BOOMING DUNES

For centuries explorers have reported hearing strange sounds emanating from sand dunes, including something similar to distant kettle drums, artillery fire, low-flying propeller aircraft, string basses or pipe organs: these are "booming dunes." The sound is caused by closely packed sand grains sliding over each other as they slip down the soft slope of a sand dune. The layer that is sliding over the top provides the energy for the sound, while the stationary layer underneath acts like a sounding board. Booming dunes have been heard in the Sahara, the Middle East, South Africa, Chile and California—indeed pretty much wherever the right conditions exist.

Find the right dune, such as Sand Mountain in Nevada, and you can make it boom all by yourself by sliding down the sand to set off the vibrations.

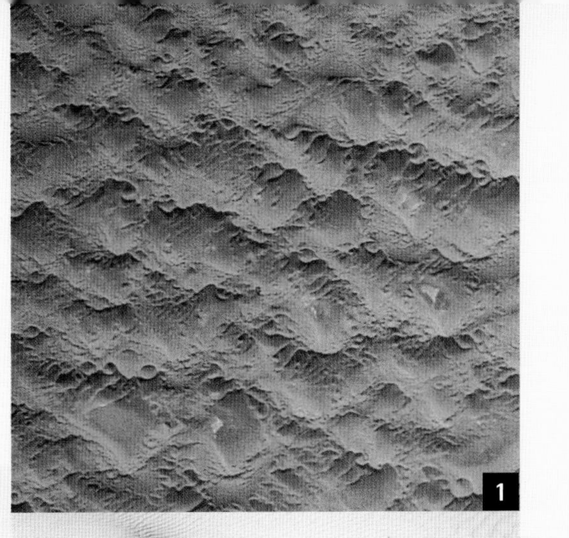

FOUR COMMON TYPES OF DUNE FORMATION

1) Star dunes in the Sahara: these form where the wind direction frequently changes, piling sand together.

2) Seifs in Death Valley, California: where there is plenty of sand, and a constant wind direction, a sharp crest is built up.

3) Barchan dune in the Namib Desert, south-west Africa: the crescent shape points to the wind, and the barchan is the fastest moving of all dunes in the desert.

4) Linear dunes in New Mexico: a strong steady wind creates very long parallel stripes of sand.

mirage

On a hot day it is easy to see the air above the ground shimmering when light is deflected as it passes through this warm layer. The shimmering is due to the fact that the density of the air is constantly varying in tiny parcels as it rises from the ground. As air warms and produces layers of different densities, so those layers refract the sunlight passing through them to a different degree. The air behaves like a lens, bending the rays of light towards colder air and altering their speed, so that the light strikes the eye from a slightly different direction to the one it has come from. Taken to extremes, the result is a mirage. In a desert, where the air near the surface of the baking ground is much hotter than the air perhaps 16 feet (5 meters) above it, light moving towards the ground from the sky above the horizon will bend up towards the cooler air and create an inverted image as it strikes your eyes. Meanwhile, light will also move directly from the same part of the sky to your eyes and appear in its normal position. So suddenly a reflected image of the sky appears below itself—and the only interpretation that the human brain can make of such a clear image is that there is a lake of blue water just before the horizon line. It is not an illusion as such, because the light has genuinely been refracted, so it is very hard to fight the impression that the shimmering image created by the reflected sky is really there. This kind of mirage is called an "inferior" mirage, and is the normal effect of the sun in a desert.

However, there is another sort of mirage, the "superior" type, which is created by the exact opposite conditions. Here the air near the surface is much colder than that slightly higher up. In that case the light is refracted in the opposite direction—light heading upwards is refracted

Above: A mirage in the Western Desert

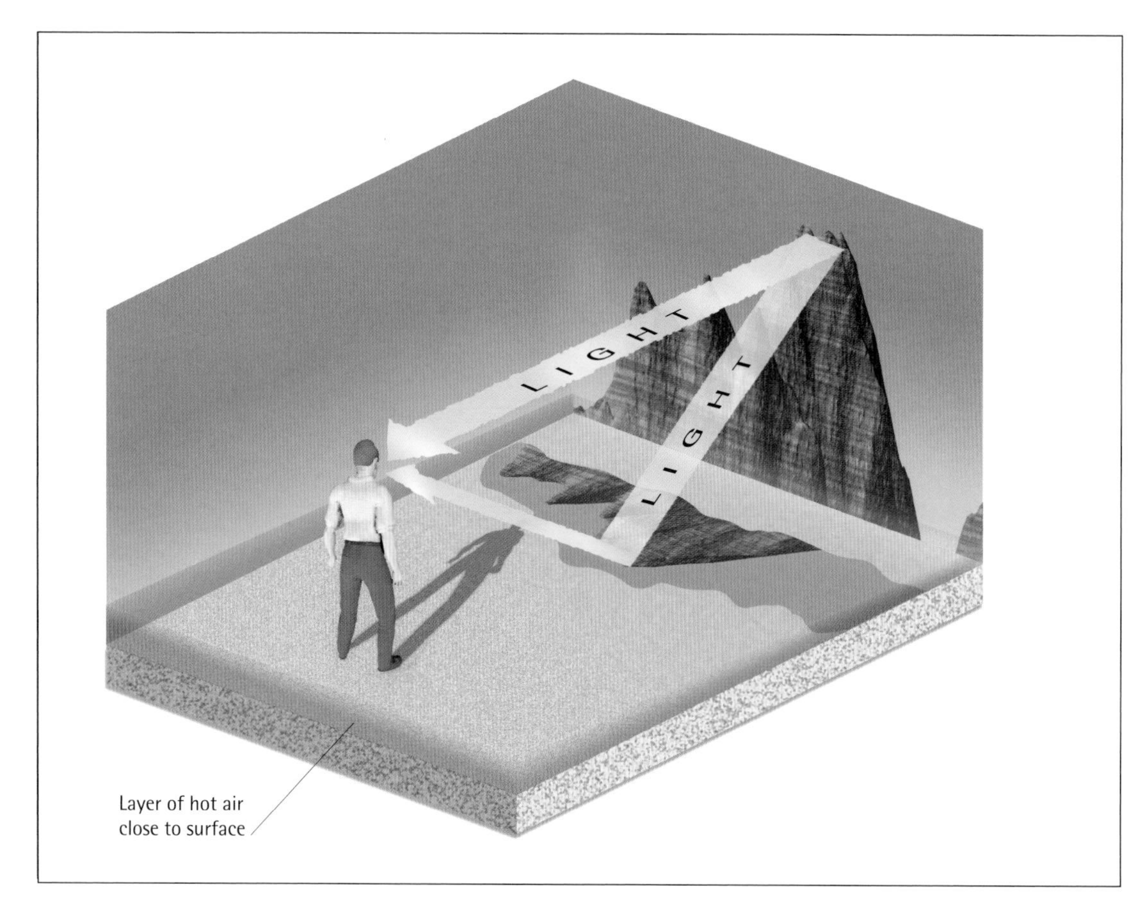

Left: The layer of warm air above a desert surface behaves like a lens, so that some of the light from a distant object is refracted to reach your eye as an inverted image from the direction of the ground. The result is a mirage—an image from apparently below the horizon of something which is not really there.

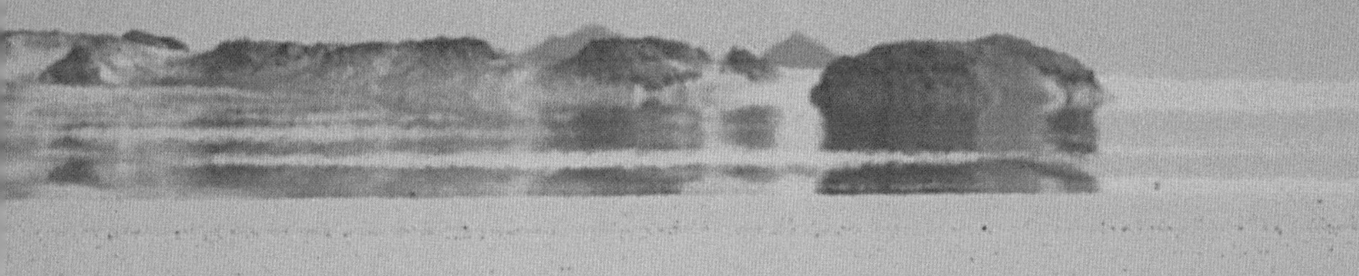

of Egypt. Light from distant mountains and the sky is refracted, to create a reflected image below the horizon line. Here this image is misinterpreted as a lake.

back down, so ultimately an inverted image is seen floating above the surface. This kind of mirage does not occur in deserts, but is common over cold oceans or wide expanses of ice where the air near the surface is made very cold. Such an image is known as a Fata Morgana, named after the magical temptress Morgan le Fay in the legends of King Arthur and Camelot. The effect of a Fata Morgana in the open ocean can be to make the image of a ship or island that is actually far beyond sight, over the horizon, appear within easy reach. Legend has it that the stories of the *Flying Dutchman*, whose ghost was doomed to sail the seas forever, stem from the ghostly images of ships captured by a Fata Morgana. And in cases where the air forms several layers of different densities, the light can be refracted several times over: the Straits of Messina, between Sicily and mainland Italy, are fabled for their fleeting visions of phantom cities, castles and towers populated by hordes of armored soldiers.

It is perhaps ironic that the optical effects of warm and cold air are such that the phantoms they create for desert explorers or sailors lost at sea are precisely what each most longs for: an oasis of water, a friendly ship or a welcoming shore. They are not only deceitful, but also cruel: the temperature and movement of the air is constantly changing in the swirling motion of our atmosphere, and these tantalizing images can appear for long enough to give hope, only to disappear, reappear and vanish just as quickly, teasing their victims in their hour of need.

Above: Negotiating a wadi: a dried river valley that can quickly flood in a sudden desert rainstorm.

desert storms

It would be wrong to say that it never rains in the desert, but doesn't rain much, and when it does, rainfall tends to be short, sharp and violent. With the approach of summer in the north, the trade winds from the southern hemisphere actually stray north across the Equator, where our old friend the Coriolis effect swings them round to the right. Off the west coast of Africa these new winds pass over the Gulf of Guinea, picking up moisture along the way, and move on towards the interior of the continent, where they meet hot, dry air sweeping down through the Sahara. As these airflows converge, the hot, moist air rises to great heights over the Sahara and creates the few spectacular thunderstorms that build over the desert.

The rains are rare, but when they come the desert reacts quite differently from other parts of the world. With little or no vegetation, and only a thin layer of dusty topsoil, the ground cannot soak up the torrential rain which quickly begins to run across the surface. The result is sudden and very large flash floods which carve channels across the landscape—the characteristic "wadis" that can be seen as steep-sided, flat, dry river-beds for most of the rest of the year. The sudden rushes of water also add to the erosion of the land. They scour across the already fractured rock and gravel of the surface, sculpting out new channels, picking up the thin layer of desert sand and soil that has loosely blown across the land and carrying vast quantities of it as sediment in their wake. In 1973 a huge storm on the Wadi Medjerdah in Tunisia created a flash flood that laid down a layer of silt and sand across an area of 53 square miles (138 square kilometers). This was an extraordinary reshaping of the desert but, to put it into perspective, it was an event that occurs perhaps once every two hundred years.

Above: A dust devil—a dust whirlwind.

Above: A building dust storm approaches.

haboob

Desert thunderstorms are spectacular but usually short-lived, lasting perhaps no more than a couple of hours. But in their wake can come something much more desert-like and much more frightening: a haboob. The name comes from the Arabic word *habb*, meaning "to blow," which is perhaps an understatment. A haboob is a large wall of dust, terrifying in its scale, which builds up at the leading edge of a desert thunderstorm. At the heart of the storm is a rising updraft of warm air and a falling downdraft of cold. As the falling, cooling air hits the ground it spreads out, away from the storm center, creating a "gust front" of wind which can be very fast, lifting dust and sand in its wake to create the dark, towering cloud of dust that is the haboob. If you could see it from above, the storm would seem like a wave breaking on the beach—turbulent, churning, scouring the ground in a maelstrom of sand, dust and gravel. The wind damage can be severe. But the haboob can build to something even bigger. In the tropical ocean, collections of thunderstorms that have formed in a squall line can find themselves binding together, to create the slowly whirling monster that becomes a hurricane. In the desert, a line of storms can in turn create a line of haboobs that slowly move out of the desert carrying huge volumes of dust before them, to become a continental-scale duststorm.

There are other kinds of large storms that emerge from Africa, events during which the seasonal winds that flow off the desert persist

Above: A severe sandstorm.

and carry billowing clouds of dust before them. Whatever their origin, these large-scale clouds of dust can move the sands of the Sahara literally across the world. To the north, Saharan dust is borne across the Mediterranean to fall as a fine layer over the holiday beaches of southern France and Italy. It may even reach Paris then move across the English Channel and up through Britain.

Today, spectacular satellite images reveal the extraordinary reach of the world's greatest desert. Along the convergence zone (the intertropical front that sits across the center of Africa and rests just north of the Equator in the summer) successive centers of low pressure move from east to west, in waves of disturbance. These "easterly waves" bring weaker then stronger winds that gather up the dust in concentrated waves, perhaps 600 miles (1,000 kilometers) wide. These move gradually across the desert, rising in the turbulent air to heights of several miles above the ground. The pictures from space clearly reveal the clouds of desert dust and sand as they move west from the African coast and over the Atlantic Ocean. They are borne along high in the troposphere, to spread out and die over the Caribbean where they quietly rain their dry contents on to the lush islands, gently adding sand to the holiday beaches that run into the tropical ocean. The scale of this continental movement is almost unimaginable. It is thought that up to a billion tonnes of dust move out of northern Africa every year—half of all the dust that is deposited in all the world's oceans.

thunder and lightning

The summer sun warms the tropical ocean, fueling the hurricane season with the warm, moist air that rises as a result. And from the Caribbean, across the Gulf of Mexico, large humid air masses reach north across the hot plains of the southern USA. Water and heat bring the muggy weather of summer and the conditions that create the massive rain clouds that routinely build across America each year. Those storms bring the most thrilling of the weather's array of special effects: the signs from the gods—thunder and lightning. Around the world, at any one time, there are probably 1,800 thunderstorms in progress and a hundred lightning strikes every second. Recently the occurrence of these strikes has been mapped across the globe using satellite imagery and it has been established that lightning follows a clear pattern. It builds up in the afternoons over land, occurs more frequently over land than over the sea and shifts with the seasons, moving north with the sun in the northern summer and shifting south for the other half of the year.

For such a common and impressive natural phenomenon, it has been extraordinarily hard to discover the exact mechanism of a lightning strike, ever since Benjamin Franklin started the process off with his famous experiment when he flew a kite in a thunderstorm. Even today there is heated debate over what exactly is going on.

Left: A computer-generated image of a lightning bolt—seen from the point of view of the thundercloud.

LIGHTNING STRIKES

Inside the body of a fully formed, mature thundercloud there occurs the extraordinary phenomenon that makes us aware of its awesome power: thunder and lightning. The cloud is a maelstrom of air currents, rapid downdrafts flowing next to equally strong updrafts, with particles of ice and hail falling and rising as they are caught in the cross-currents. In their travels through the downdrafts, ice particles that have formed higher up in the icy region of the cloud collide with water droplets, which instantly freeze. This freezing releases heat, keeping the surface of the ice particles warm. When a warmer ice particle collides with a smaller, cooler ice crystal, positive ions from the warmer object are transferred to the colder object. The result is that the tiny, colder particle is now positively charged and the larger, warmer particle is negatively charged. The tiny particles are lighter and so are carried up by the updraft, taking their positive charge with them, and the top of the cloud becomes positively charged overall. Meanwhile the warmer, larger, heavier, negatively-charged particles have been carried further downwards in the downdraft, making the bottom of the cloud negatively charged.

As this negatively-charged cloud moves slowly across the Earth, there builds up on the surface of the ground a positive charge (because opposite charges are drawn to each other), which tracks

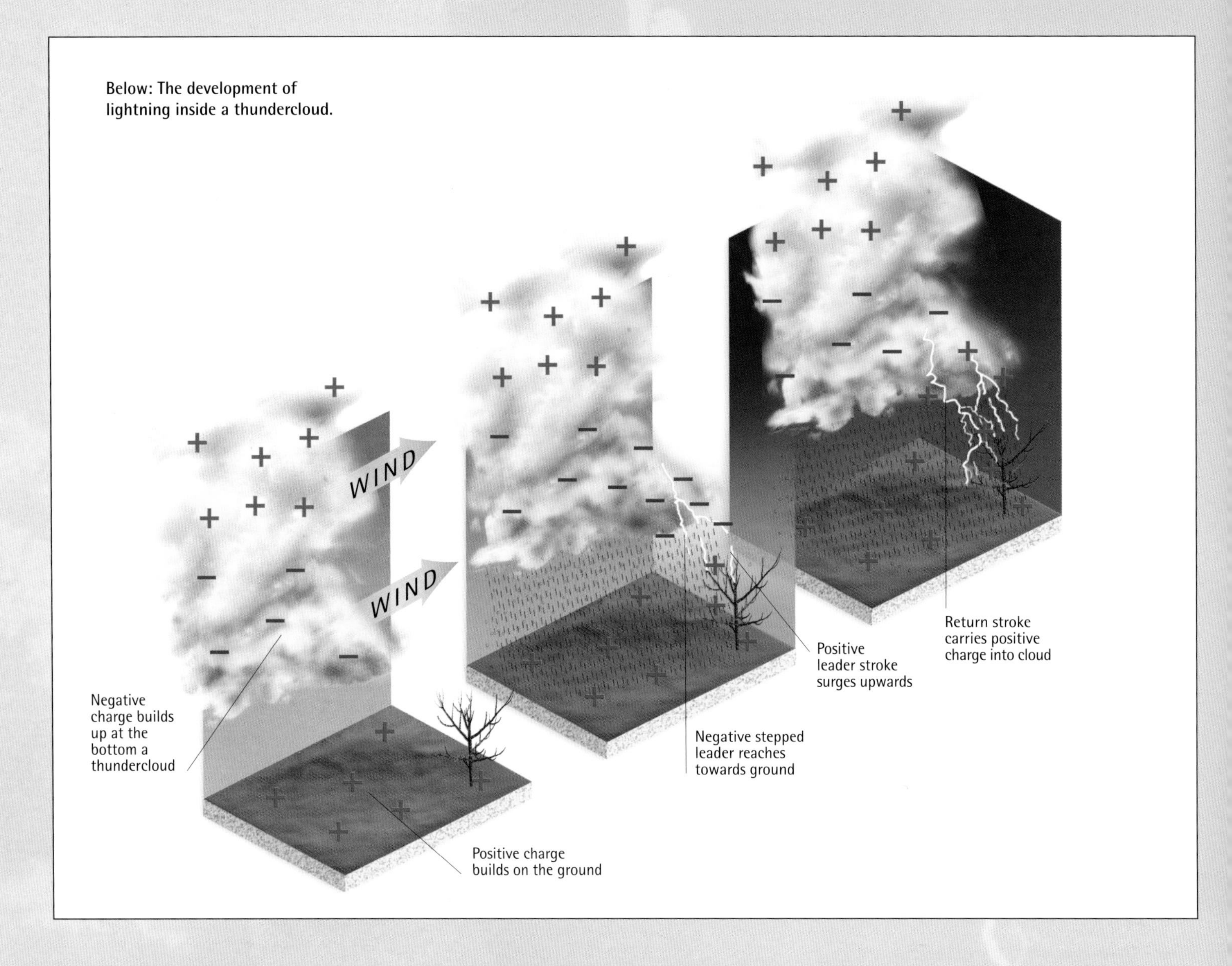

Below: The development of lightning inside a thundercloud.

Above left: The bright visible stroke of lightning is in fact the "return" path of the positive charge moving upwards into the cloud.
Above right: A double lightning strike from a silver-lined cloud in Arizona.

along below the cloud. The difference in charges between the cloud and the ground creates a huge electric potential, but as the air is a very good insulator these two opposites at first cannot meet. However, as the charge builds and builds in the cloud, the electric potential gradient between it and the ground increases to something like a million volts 3 feet (1 meter), and a point is reached when the air cannot hold back the negative electrons any longer. A surge of electrons rushes down to the cloud base and on towards the ground, but it does not flow continuously. Instead it moves in a succession of steps: the charge rushes down at a speed of about 14,000 miles (220,000 kilometers) per hour for about 150 to 300 feet (50 to 100 meters), then stops for perhaps a fifty-millionth of a second, then travels on again for another 165 feet (50 meters), and so on towards the ground. This halting progress is known as the "stepped leader," but it is so fast that it is invisible to the human eye. As the negative charge begins to get near the positively charged ground, positive ions move up from the ground through any protruding object such as a tree, a tower or even a human. When they both meet a strong electric current flows, carrying positive ions up into the cloud along a channel several inches wide. It is this "return stroke" that glows brightly enough to be seen. It moves up from the ground like a giant spark, but so fast that its direction of travel cannot be noticed. A typical lightning flash will consist of about four leaders and return strokes. But it does not stop there. Lightning is incredibly hot—at a temperature of some 86,000°F, it is five times hotter than the surface of the sun—and this extreme heat causes the surrounding air to expand explosively, creating a shock wave that travels through the sky as the sound of thunder.

You can calculate how far away a thunderstorm is by timing the gap between seeing a lightning flash and hearing the sound of the thunderclap. This is because sound is actually very slow-moving—at approximately 683 miles (1,100 kilometers) per hour—compared to the speed of light. Every three seconds before you hear the clap means that the storm is ½ mile (a kilometer or so) further away. But watch out: even within 6 miles (10 kilometers) you are in the danger zone for a lightning strike!

fossil lightning

When lightning strikes a human body the effects are swift, powerful and also insidiously long-lasting. A typical strike delivers about 300,000 volts in just a few milliseconds, and the bulk of the electrical charge passes over the surface of the body in what is called an "external flashover." But at the points of contact, usually the head, neck and shoulders, there will be deep burns where most of the physical tissue damage is done. The immediate risk is total cardiac arrest but if you survive, or are fortunate enough to be resuscitated quickly, the problems don't end there. Neurological damage can often emerge much later: amnesia, seizures, motor-control damage, hearing loss, blindness, sleep disorders, spinal-cord injury and paralysis. Lightning is not a phenomenon to take lightly. After rain, cold and cyclones, it is the fourth most deadly weather-related killer.

Remarkably, lightning can also leave its trace in the ground itself, and in areas where lightning is common, such as Florida, you can easily come across fossilized lightning strikes. Known as fulgarites, these form when a lightning bolt burns into the ground, melting the minerals in the soil as it passes so that, as they cool down, they form long, hollow, glassy tubes. They are to be found among the sands of many beaches, broken and worn away. But recently researchers in Florida, where lightning is dangerously frequent in summer, stumbled on a fulgarite that was over 16½ feet (5 meters) long and branched into three forks, driving its way down into the ground. It is now in *The Guinness Book of Records*.

Above: A fulgarite, the glassy remains of minerals melted by lightning striking the ground.

summertime hell

Even away from the extremes of the deserts, heat is a force that can take humans by surprise and have deadly effects ●

The heat that moves gently north with the sun over the continental interiors of the northern hemisphere brings summer and all that goes with it: thunder and lightning, but also a general sense of life getting better. Yet even away from the extremes of the deserts, heat is a force that can take humans by surprise and have deadly effects. In July 1995 an unprecedented set of factors conspired to bring five days of hell to the city of Chicago, leaving over seven hundred dead.

At the start meteorologists had no real sense of what was to unfold. They could see that a heatwave was emerging, but that was nothing unusual. "Although we forecast there would be a heatwave," said Paul Dailey, a senior meteorologist on duty during the event, "we didn't realize the seriousness of how it would affect people, and we didn't think of it as a killer. It caught us by surprise when we saw them carting out bodies from the houses." What made this particular summer weather so deadly was a unique combination of events that came together. It had been raining for several days across Iowa and Missouri, and the ground was saturated with water. On the great fields of corn that spread out across the plains of those Midwest states, the farmers of America's breadbasket were relaxed in anticipation of a bumper harvest. The corn was growing well under the perfect combination of strong sun and plenty of water. In the wake of the rain an area of high pressure had formed over the region. This was not unusual for the time of year, but this one had been amplified by a strong disturbance out in the Pacific Ocean some days before and was therefore particularly intense. It produced bright clear blue skies; the summer sun, now near its most northerly position for the year, quickly warmed the air and the ground, and the people of Chicago headed for the shores of the Great Lakes, fresh from the great American winding-down of the July 4 holiday weekend. But high above the first signs of trouble were brewing: the jet stream, the high-level, fast-moving wind that flows around the planet, was particularly weak, and had redirected its path up to the north. As a result there was little high-level wind to shift the high-pressure system and it stuck, sitting stagnant over Chicago, quietly, relentlessly allowing the sun to heat the city and the surrounding region.

In the open plains of the grain belt, over which the air mass had passed on its way to the edge of the Great Lakes, it was now the height of the growing season and, as the corn was baked by the sun, it rapidly drew up water from the damp soil and transpired it quickly from its leaves. Huge volumes of water are lost from the ground through plants in this way each day, and as the air mass moved slowly overhead it picked up water vapor and was very humid when it arrived at Chicago. For the people of the city and its outskirts that alone would have been nothing too unusual, but there was another factor: a huge tongue of warm, moist air had begun to flow up from the Gulf of Mexico. This added tonne after tonne of water vapor to the air—creating even more humidity. The humidity built up to over 90 per cent in the late evening and the moisture in the air created an inversion, a layer of air so warm that it prevented the air below from rising. The heat was trapped below. But the inversion also trapped something more directly harmful. The exhaust fumes from the city's vehicles and industry were unable to rise and became held in a huge bubble of pollution that hung over the urban areas.

The uniform high pressure that had settled over the city meant a lack of wind, so people could find no relief from the intense heat and stillness, which was only added to by the effect

Above: Thick heat-haze and smog over New York in summer.

of the structure of the city itself. Downtown Chicago contains some of the tallest skyscrapers in the world and in these conditions, capped by the temperature inversion, the hot air formed in ponds between the buildings. It was almost glued to the ground, unable to circulate and able to do only one thing—get warmer. And the city itself acted as a bakehouse. The vast areas of concrete and asphalt that make up the modern conurbation of Chicago's southside absorbed huge amounts of heat, turning houses into ovens with temperatures reaching 104°F (40°C) indoors at night. Some people even tried to put shiny material on their roofs to reflect sunlight, in a desperate attempt to stay cool.

Perhaps the final straw in all this remarkable combination of factors was the behavior of the people themselves. Heatwaves have always occurred in this region, but fifty or sixty years ago people would leave their homes and whole families would sleep outside in a park to get what fresh and cool air they could. Today, however, the people of an inner city could not conceive of sleeping out at night. In Chicago all the doors were locked and the curtains drawn, and every citizen shut themselves away—in an oven of their own making.

Over the three days of July 13 to 15 seventy temperature records were set across America, from the Great Plains to the Atlantic coast, but in Chicago and Milwaukee records were also set for numbers of victims. Compared to the violent, destructive and obvious nature of hurricanes, tornadoes and floods, which are regarded as our number one climatic enemies, heat—just like extreme cold with its insidious effects—is a silent killer. In particular, it is hard to know when the danger has really begun, for the onset is slow and by the time people are aware of what has befallen them it may already be too late. Heatwaves and deaths resulting from them are not unfamiliar to cities like Chicago and on July 12, 1995, as the forecasters were issuing warnings of temperatures nearing 104°F (40°C) the next day, few people were overly concerned. Chicago is a violent city, and the first signs of the natural emergency was when the police began to report a sudden increase in "non-violent deaths" during the course of the 13th and 14th. By the time the authorities gave a severe health warning on July 15 it was already the last day of the wave of record temperatures that had engulfed the city. On that day alone 162 heat-related deaths occurred, but by then many Chicago hospitals were already on "emergency bypass," turning away new victims because their emergency rooms were full. As in so many natural disasters it was the elderly who suffered most, particularly those living alone who were unable, or perhaps unwilling in the city's threatening social atmosphere, to seek help outside their baking homes.

heat island

The effect that Chicago itself had on its own weather during the heatwave of 1995 was extreme and deadly, but it was only one indication of how the modern world can influence the weather. The notion of global warming is now part of our twenty-first-century lexicon, but the idea that local weather is produced by human activity is something that we have only recently faced up to. Have you ever wondered why the weekend never turns out to be as good as you'd thought? There you are, all through the week, working hard in the office or factory while the sun blazes outside. It makes you long for that weekend break when you can lie out in the garden, cool glass in your hand, book dropped carelessly on the ground as you build up a tan—or perhaps take a trip out of town to enjoy the sunshine in the countryside. Yet when the weekend comes Saturday dawns with scudding clouds and builds to a nice damp afternoon; and as you pack the picnic hamper on Sunday morning, listening to the reports of rain in the countryside, you decide that maybe you'll just stay in, in front of the TV. Why does that always seem to happen? There are the beginnings of an answer, and it seems that it could well be your own fault.

As usual the effect first became clear enough to be studied in the USA, where the growth of large conurbations in the last few decades has set meteorologists wondering about the influence of city dwellers on the local climate. The city of Atlanta, Georgia, is a huge urban sprawl that is getting bigger. In the last thirty years almost 580 square miles (1,500 square kilometers) of forest have been cleared from around the city to allow for urban growth. The result is an area of concrete, asphalt, steel and glass that quite simply soaks up heat.

Atlanta is not unusual: the idea of the urban "heat island" was first proposed by an English amateur meteorologist, Luke Howard, in 1818 after ten years spent measuring the temperature in London and comparing it with the surrounding countryside. However, it is in America that it is seen at its most extreme. For example, temperature records in Los Angeles go back far enough to provide a comparison with those before the rapid post-war urban growth in the USA. In LA summertime high temperatures in the 1930s never rose above 99°F (37°C). Today they regularly exceed 104°F (40°C). And city temperatures are far higher than those in the surrounding countryside, by at least 9°F (5°C). For years now it has been recognized that the effects of urban heat and the pollution produced by cars and industry increase levels of smog over the typical city, but it has emerged in the last decade, from studies of Atlanta and New York, that the city creates its own weather and even changes the weather over the surrounding countryside.

As the heat builds up over the urban concrete jungle, so the warm air rises above it and diverges aloft, creating low pressure over the city. From the surrounding areas, cooler air is sucked in to replace that which is rising. In effect what is happening is the same cycle that builds towards a typical summer thunderstorm. And thunderstorms are precisely what have been found to occur over Atlanta and New York. On calm days the storms form over the city itself: in New York they build up towards the late afternoon; in Atlanta, as satellite images of the weather patterns across the region clearly show, storms suddenly appear as if out of nowhere amid totally clear skies across the southern states, and dump heavy rain on to the unsuspecting citizens below. But it seems that the modern city has another surprising effect, one which showed up in New York which has such a large number of very tall buildings at its center. On rainy days the skyscrapers of the Big Apple disappear well into the cloud, so perhaps it's not surprising that they have an effect on what happens up there. On days when there is a general flow of wind across the region, and when the city is hot enough to generate its own storms, the buildings and the rising column of warm air act like a barrier to the prevailing winds. Instead of forming over the city, the storms that the heat island builds up therefore work their way around its edges to dump their rain on the countryside downwind of the conurbation. So if you move to the suburbs of New York, it is probably worth finding out where the generally prevailing winds come from in summer and buying a house upwind of the city.

But it doesn't stop there. There's that question of the weekend weather. Remarkably two scientists from Arizona, Randy Cerveny and Robert Balling, who studied all the weather-satellite data that had been gathered for the east coast of the USA over the last twenty years were amazed to find that on the coast the weather really is wetter at the weekend than during the week. The notion of weekday and weekend is entirely a human one—nature doesn't take time off—so if the weekends are wetter it must be down to something we humans are doing. And the key seems to be city pollution. Just a simple graph of pollution levels in a city reveals that they build up during the week as people go about their atmospherically dirty business from Monday to Friday. Thursday is always worst, which probably confirms that people start to slope off early the following day with that "Thank God it's Friday" feeling. Pollution build-up leads to more heat being trapped over the city. Also, particles of air pollution can themselves act as seeds for cloud droplets to form, so more local rainstorms are created. Meanwhile, the prevailing winds across the USA are westerly, so the city-generated rain and cloud will always tend to be blown out to the coast. At this stage there is no better explanation, but the effect is a real one. And the same two scientists also found a truly astonishing further pattern in the weather. Out at sea it rains more at weekends too, but the effect on tropical storms and hurricanes is quite the reverse. Believe it or not, Atlantic hurricanes are less severe on weekend days! Again there is no hard evidence as to how this would work, but one theory relates to atmospheric chemistry. It is that when the large amounts of carbon contained in city pollutants are dumped on to the middle of a hurricane they could cause the storm to spread out, thus widening the eye at the center and weakening the hurricane. So when you head out for the beach on Sunday you can be reassured that, while it might rain a bit, you are less likely to be blown away.

Below: Heavy smogs forming over Los Angeles have prompted legislation to restrict cars by 2011.

cautionary tale

The unexpected and unwanted effects of a modern, human-created heat island are literally and perhaps most starkly thrown into focus by the bittersweet story of the remote land of Nauru, whose tale is a cautionary one for all humanity. Nauru is a tiny coral island in the western central Pacific Ocean. For aeons it was a favored resting point for the great and gracious frigate birds that glide the length and breadth of that ocean. These remarkable creatures can fly thousands of miles across open sea, gliding for days at a time, borne aloft by their vast wingspan, swooping low to fish with their huge hooked beaks, sleeping on the wing, navigating with a sure touch from tiny island to tiny island, sometimes not even bothering to stop. Nauru was a favorite spot for them to rest and breed, and the island's limestone rock gradually build up a depth of bird droppings that was perhaps unlike any on the planet. Nauru was deep in guano: gullies, hollows and valleys between its outcrops of coral were all filled with bird excrement, which provided a rich soil on which grew coconut palm, wild cherries, figs, breadfruit and so on. For the Polynesian people who colonized the island, probably thousands of years ago, it was a lush tropical home and the frigate bird became their emblem. When the first European whaling ship stumbled across it in 1798 its captain named it Pleasant Island, for all the idyllic reasons that one would expect.

Nauru and its bird droppings would have probably remained unbothered and unchanged had it not been for the sheer chance of one Arthur Ellis tripping on a doorstop in an office in Sydney in 1900. Ellis was a New Zealander who was working for a London mining company, and the doorstop was thought to be a rather fine piece of fossilized wood. But, being a chemist, he decided to test its composition and, to his delight, it turned out to be of the finest quality phosphate, perhaps the best in the world. Phosphate rock is the finest possible fertilizer; it comes from ancient deposits of guano; the doorstop turned out to have come from Nauru.

Above: Mining of phosphate from guano on the island of Nauru.

Thus began a century of phosphate mining on Nauru, digging into the lush center of the island—the higher ground the islanders know as "Topside"—which supplied them with so many of their crops and timber and was home to the beloved birds and wildlife that shared their island. Within the first five years over half a million tonnes were shipped away, to a value of almost $1.5 million—and that was before the First World War. The Nauruans received a tiny royalty for each tonne—mere pennies—but, even so, they were to become fabulously rich. So it continued throughout the twentieth century. After passing through German, British, Japanese and Australian hands, the island regained its independence in 1968. It became the world's smallest republic, and its citizens found themselves with the highest income per capita in the world. By the 1990s the mining royalty trust fund for the islanders was worth $730 million. Health care and education was for all, along with cars, air conditioning, electronics and Coca-Cola.

But the price was heavy; more and more of Topside was removed, year on year, the trees cut down, the soil stripped away, the land left as bare rock. While the islanders had more cars and

motorbikes per head than those of any other Pacific island, the road ran only for 12 miles (20 kilometers) along the coast. While they imported food and building materials for a better lifestyle, the vegetation of the island was reduced to a thin, green strip that runs in a ring around the shore. To visit Topside today is to gaze out over a bleak, gray, lunar landscape, not a tree, not a shrub, not a fleck of green or color; the island's birdlife is almost gone, the air is silent but for the clanking and groaning of the last remaining mining equipment. The bare coral, exposed after millions of years of natural cover, now stands out in craggy pinnacles with soil hollowed out from between them, and a fine layer of acidic phosphate dust wafts gently over the island. The heat is intense.

Nauru's weather has changed. The heat that rises from the surface of Topside blasts an updraft of warm air high into the Pacific sky, so fast and so strongly that it disperses any cloud formation above the island. And because the air rising from the bare rock is devoid of moisture, there is nothing to condense as new cloud. Rainfall has been significantly reduced and the little greenery that remains is suffering. Palms stand dead, the crowns of pandanus trees are scorched and even the prized golf course has no grass left. The country has experienced three successive years of drought—on a Pacific island where humidity, cloud, rain and sea breezes should be a way of life.

Nauruans have suddenly woken to the realization of what has happened to their home. Of their fabled wealth there is sadly little left, for as the century turns the island is now effectively mined out and their income is no more. The trust fund dwindled to a mere $146 million in ten years after some disastrous financial decisions: the money-losing airline, the London musical *Leonardo* that flopped, innocent investment into crooked financial scams. In a last-ditch attempt at recovery Nauru has sued her former colonial masters for compensation for lost mining revenue, and has reached a settlement which the government hopes to use to rehabilitate the island. Over the next twenty years a team of scientists—biologists, geologists, ecologists—has a plan to level the pinnacles of Topside, import soil from neighboring countries, sterilize it against unwanted microbes and regenerate the heartland of the island, replanting it with the traditional Polynesian plants and trees and reintroducing the traditional birdlife. Perhaps, they hope, even the mighty frigate bird will return.

But Nauru has little time. The tragic story of the island that consumed itself is a lesson for us all, a story of the world in microcosm as we face the effects of the heat we have created across the planet as a whole. Nauru is also caught up in that global picture, for it is one of the many tiny low-lying islands that are dotted over the tropical oceans. The thin strip of populated land that surrounds the desolate Topside is barely 3 feet above the waves that surround it and is vulnerable to tidal surges and flooding. It is ironic that Nauru will be in the first wave of likely victims of global warming as the level of the oceans inexorably rises in the century ahead—when not just this island's climate will change, but so will that of the whole world.

It is to the changing of the weather that the final chapter of this book turns.

Humans have long sought to control the weather, but the consequences have rarely been foreseen. The first scientific attempts to make rain brought only lawsuits, and one scheme intended to improve the weather would have risked the onset of an Ice Age. But now the climate really is changing under our influence, and the results may be more sudden and more violent than we ever could have predicted.

chapter six

change

Above: A hurricane-driven wave smashes into an American coastal sea-wall.

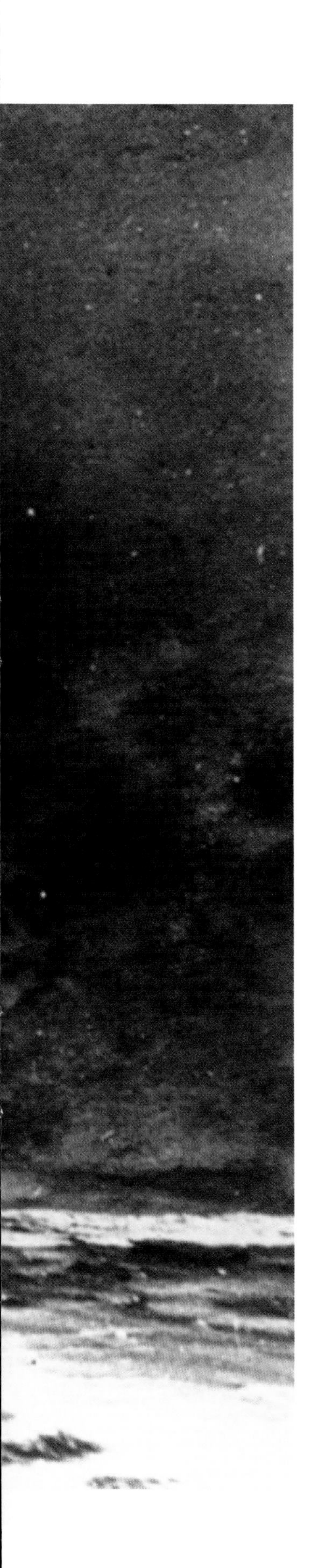

meddling with nature

It was not the worst hurricane of the year, but, as the end of the hurricane season drew near, the residents of Miami knew that it was certainly the second worst. On October 13, 1947 the storm had moved away from the city leaving a trail of destruction, and had begun to turn out to sea. It was following the path of so many hurricanes, wheeling around the subtropical high-pressure zone and heading north-eastwards up the coast of America. On that day its center was some 300 miles (500 kilometers) offshore and, if anything, its energy was beginning to abate. But on that day a B-17 bomber made three flights into the maelstrom, each time buffeted by the terrifying winds, and each time unloading 330 lb (50 kilograms) of solid carbon dioxide, "dry ice," across the path of the storm. It was the first test of Project Cirrus, the ultimate dream of weather scientists: a scheme to control the power of a hurricane. As the aircraft returned to its base the scientists proudly announced that the cloud had clearly reacted to their treatment, but as the day wore on they were disappointed to find that their action had not caused the storm to die. The next day, however, their disappointment turned to horror, as the hurricane turned back on itself, reintensified and swept into Georgia, causing a new wave of destruction and death.

taming the weather

The events of October 1947 set back the cause of weather modification by some years; lawsuits for damages were filed against the experimenters and only dropped when records were dredged up that showed a similar hurricane forty years before had made an identical volte-face entirely of its own volition. But if their fingers were burned, so to speak, the scientists who wanted to tame the weather did not give up. They were following in a great tradition and had a worthy goal—to make the world a better place for humans to live in.

The distant history of weather modification is riddled with superstition and religious faith. Thunder and lightning, tempests, flood and drought have long been ascribed to the work of the gods in all cultures. Prayers for rain exist in every society—the Mayans sacrificed middle-aged women to call for it; less murderous societies would settle for the manipulation of animal entrails; the Mongolians poured milk on to the ground; others merely sang or danced. But all of them had the aim of making the weather, or rather the gods that controlled it, do what they wanted. Attempts at a more reasoned attitude to cause and effect came only in the last few centuries. Church bell-ringers in

Above: A Mayan rain god.

Above: Irving Langmuir (1881–1957) won a Nobel prize for chemistry in 1932, and went on to become a pioneer of weather modification.

eighteenth-century England peeled their bells so that the noise would shake rain from the sky, while the soldiers of the American Civil War were convinced that firing shells into the air brought heavy rainfall after their battles—and, indeed, early experiments to create rain in Texas centered on dynamite explosions. At the turn of the twentieth century, French wine-growers more naively thought they could limit hailstorms, which have long been the bane of vineyard owners in south-western France, by firing anti-hail cannon into the air. They believed that the gunpowder would literally shatter the hailstones and turn them into gentle drops of rain. Needless to say, it didn't work.

During the Second World War the pressures of wartime forced the pace in weather modification, and one generally accepted success was fog dispersal. British bombers regularly experienced difficulty landing in fog on their return from raids across Europe, so the Air Ministry set up the Fog Investigation and Dispersal Operation (FIDO). The technique was simple, if a little excessive. Fog forms when very moist air cools and some of the vapor in it condenses into cloud droplets close to the ground, so the British military went straight to the root of the problem. They set up an array of pipes alongside the aircraft runway through which gasoline was pumped into the air, and burnt to heat the air and evaporate the fog droplets. The project was a resounding success, and some two thousand aircraft landings and take-offs were aided in this way in the last two years of the war. The cost, however, was astronomical: over 15 million gallons (68 million liters) of gasoline and oil were consumed, burning money at a rate of anything from $150–$450 per minute.

Post-war science got really serious, however, culminating in the remarkable Project Cirrus. It all started with two men, Irving Langmuir and Vincent Schaefer. They worked in a laboratory of the giant General Electric Company, but during the latter part of the war they spent two winters at Mount Washington, enduring the extreme weather conditions at the research station there (see Chapter Four) while they tried to understand the formation of rime ice in order to assist in preventing its build-up on aircraft wings. At Mount Washington they became intrigued as to why a snowstorm would suddenly form from a cloud that had apparently been content to remain as a collection of supercooled droplets of water. They wanted to understand the mechanism whereby nuclei in the air trigger the formation of ice crystals, the process we know today to be at the heart of the formation of both rain and snow. However, in the course of their quest they hit upon something altogether more awesome in its implications. Back at their company laboratory, Schaefer purchased an ordinary domestic freezer (GEC of course) in which to do experiments. He tried to create ice on particles of soot, graphite, volcanic dust, sugar, talcum powder, salt and soap, all to no avail. But one hot afternoon, when the freezer simply would not cool far enough, he dropped in a chunk of dry ice—solid carbon dioxide, which rests at a viciously cold temperature of -109°F (-78.5°C)—to help it along. To his astonishment and delight, a flurry of ice crystals formed inside the freezer, grew larger and fell as a tiny snowstorm! We now know that at the critical temperature of -40°F (-40°C) supercooled water begins to turn into ice crystals, which themselves become the seeds for snowflakes. For Schaefer and Langmuir, the result meant that they could see how to control the weather.

Their first experiment "in the wild" was carried out in November 1946, when Schaefer seeded 6½ lb (3 kilograms) of dry-ice pellets across a 4-mile (6-kilometer) stretch of supercooled altocumulus cloud 15,000 feet (4,500 meters) above New York State. The cloud immediately became disturbed and within minutes snow was falling. The experiments continued, but immediately after the fourth trial the heaviest snows of that winter set in. The GEC lawyers began to worry that they might be held responsible, so the experiments were curtailed to the laboratory once more. There another colleague came up with the idea of forgetting about the tricky business of forming ice crystals themselves and instead seeding the cloud with another substance altogether—one that had a similar crystalline structure to ice and onto which the water vapor surrounding the droplets would be fooled into binding. The substance that he came up with was silver iodide, and it has been at the heart of "cloud seeding" almost ever since. The first experiments were spectacularly successful, and Irving Langmuir calculated that 220 lb (100 kilograms) of silver iodide would be enough to seed the atmosphere of the planet.

Langmuir and Schaefer were retained as consultants to the US Office of Naval Research with a plan to restart the field experiments, an event that brings the word "hubris" to mind. Project Cirrus was born amid a belief that, as Langmuir was later quoted as saying, "within one

Above: The fog dispersal system, FIDO, being ignited for a test at Blackbushe in England during World War Two.

Above: A Russian M-17 high altitude aircraft used to "seed" rain from thunderclouds during a "hail-suppression" program.

or two years man will be able to abolish most damage effects from hurricanes." There is no doubt that the scientific insights they achieved into the formation of clouds and rain were remarkable, but the implications of that Project Cirrus experiment in 1947—the hurricane that turned back on itself, and the lawsuits that hovered momentarily over the project—have dogged weather modification for all its days. The difficulty is that it is almost impossible to prove its success (or failure), because there is no way of knowing what the clouds or storms would have done had they been left alone.

stormfury

Official weather modification trials were effectively grounded for a decade after the Cirrus debacle, until the US Weather Bureau and US Navy's creation of Project Stormfury in 1962. Following wartime achievements in science—the invention of radar and the atomic bomb—after the war scientists felt that they could do almost anything. To redirect a hurricane would surely be possible, perhaps by bombing its center to break it up. But the energy in a hurricane was the equivalent of some fifty Nagasaki-type atom bombs going off every second. The approach would have to be more subtle. The Stormfury scientists came up with the idea of seeding the outer edge of the eye wall of the hurricane so that water droplets turned to ice, releasing vast amounts of latent heat. This would warm the air and reduce the temperature difference across the wall, thus raising the atmospheric pressure from the terrifying low at the heart of the storm and allowing the high-speed winds to slacken off. In August 1963 ten aircraft duly flew into Hurricane Beulah and dropped silver iodide into the clouds. The hurricane weakened. The researchers were elated. Two years later Hurricane Betsy was about to be experimented on when it suddenly changed course and headed for land. The seeding was canceled, but the officials forgot to tell the press and for months Project Stormfury was blamed for the damage caused by the hurricanes. Back on track again in 1969, the project achieved spectacular results over five days with Hurricane Debbie: the winds dropped with seeding, built up again without it and dropped again after more seeding.

military weather

In the end the project became dogged by politics rather than scientific failure. Fidel Castro spoke out at length, blaming Stormfury's experiments for the destruction wreaked in Cuba by Hurricane Florence in 1962, and complained again a few years later when a massive drought beset his tiny country. When Stormfury shifted its base of operations to Guam in the Pacific, with a view to protecting the US military bases there, the People's Republic of China objected strongly to anything that might redirect a typhoon towards their cities. At the same time the Japanese pointed out that much of their annual rainfall stemmed from tropical typhoons, and politely requested that the Americans leave nature well alone.

Above: A U S Weather Bureau team and their DC–6 aircraft, during Project Stormfury.

Elsewhere, seeding clouds simply to make rain moved ahead throughout the heady years of the 1960s and 1970s, but one of the more colorful examples again showed the grand scale of American military thinking. Operation Popeye was the name given to an audacious plan to bog down the Viet Cong as they attempted to move south along the Ho Chi Minh Trail in Vietnam. The work that led up to the operation had been partly motivated by the appalling loss of life in the Battle of the Bulge near the end of the Second World War, when Allied forces were crushed by a surprise German Panzer offensive. The German attack had gone virtually unchallenged because thick fog had settled in over the Ardennes and no Allied air cover could be mobilized. Years later in America a young scientist, Pierre Saint-Amand, moved by the tragedy, became convinced that fog-dispersal techniques could have been applied in the Ardennes, saving many lives, and so he began to work with the US Navy to perfect ways of controlling nucleation (seeding).

Meanwhile, in Vietnam, the US Army had noticed how effective the monsoon rains were in preventing North Vietnamese troop movements and decided to adopt the extraordinary idea of extending the monsoon! Saint-Amand was asked to test "rainfall augmentation" over Laos in 1966. He recalls how, when he arrived in Danang in October, the operations center was set up inside a flying C-130 of the Air Weather Service, with jet fighters from the marines doing the seeding from wing-mounted canisters, while US Army Air Cavalry planes flew beneath the clouds to measure any rainfall. This first trial was a carefully controlled set of experiments, with some clouds being seeded and near-identical ones left untouched as a control. "The seeded clouds produced more rain, grew faster and lasted longer than the control clouds," Saint-Amand remembers. "By the end of the first month the streams below were in flood stage. The trail watchers below reported that traffic along the trail had been drastically reduced." Whether or not it really worked, Operation Popeye continued to be played out for many years over the skies of Vietnam, until the US president decided to cease the bombing of North Vietnam—the military could find no category, other than "bombing," in which to place the seeding flights so they also ceased.

weather politics

As well as being used over the drought-ridden states of the USA, cloud seeding has been adopted in Russia, China, Mexico, South Africa, France and by the Bureau of Royal Rainmaking in Thailand. And there is one very parched region of the world where it has quietly been claiming success: the Middle East. Since 1961 Israeli scientists at the Hebrew University in Jerusalem have been conducting a series of programs to seed clouds over their arid lands. They have consistently claimed a success rate of some 15 per cent higher rainfall in seeded clouds. In recent years the work has been challenged, and it has also thrown up another, politically difficult, issue. The area of Israel is small, and neighboring Jordan is equally arid and equally needful of water. There are fears that a "water war" could spring up if it was claimed that a cloud that was seeded over Israel could have provided much-needed rainfall across the border in Jordan if it had been left alone.

Arguments such as this have been at the heart of disputes between weather modifiers and farmers throughout the history of cloud seeding in the southern American states of Texas and Kansas, with claim and counterclaim that, "Someone has stolen our water."

HAIL SUPPRESSION

The storms that sweep in from the Atlantic Ocean, across the Bay of Biscay, and which pour down on south-west France every spring and summer, carry with them a deadly enemy of the vineyards below. Hail causes massive damage to the wine grapes of Bordeaux each year. Hailstones are the product of the largest and most violent of thunderstorms, but they begin their life in the same way as rain or snow—as ice crystals at the furthest heights of a cloud. Once an ice crystal has formed on an ice nucleus (a microscopic particle of perhaps clay, dust, pollen or salt borne far up in the sky) high in a cumulonimbus cloud, it may stay aloft for some time and co-exist with tiny droplets of supercooled water (water which remains liquid at temperatures below zero degrees). The ice crystals grow rapidly as the supercooled water freezes to them, making larger crystals. They then become heavy enough to fall and collide with more supercooled water, which sticks to the ice crystals to form larger icy lumps called "graupel." Most graupel particles fall further, melting along the way to become rain. But some can get caught in a violent updraft of the growing cloud and be lifted up to freeze even further, become attached to more supercooled water, grow even larger and fall again. This time they may have accreted enough ice to remain frozen throughout their journey to the ground, when they will strike as hail. To become a golf-ball-sized hailstone, a graupel particle must remain aloft in the freezing part of the cloud for something like ten minutes, being caught and recaught in a repeated cycle of updrafts. Sometimes these updrafts can be so severe that hailstones are thrown right out of the top of a cloud. Pilots have often reported hail falling through completely clear air at a distance from a thundercloud.

If you cut a cross-section through a hailstone, it is possible to see the layers, like layers of an onion, where it has frozen, melted as it began to fall, refrozen with more water as it lifted, remelted, refrozen, remelted, refrozen and so on. This repeated process can produce lumps of ice of an enormous size: the largest known reported hailstone, weighing 11 lb (5 kilograms), fell on the Guangxi region of China on May 1, 1986.

The idea behind seeding clouds, to prevent major hailstorms is to seed so many ice crystals (and condense out so

much of the available water as individual droplets at pretty much the same time) that no one crystal can find enough available water to grow to a large enough size to make hail. Instead the rainfall increases. Evidence for silver-iodide crystals seeding water droplets effectively has not been totally convincing, but recently a new technique has proved more successful. This uses hygroscopic (water-seeking) salts which draw water to themselves like airborne salt crystals that naturally start the process of cloud formation. The effectiveness of the new approach was discovered by accident, when a South African weather-researcher was flying over a paper mill and suddenly found himself in a severe rainstorm where a rainstorm should not have been. Enquiries revealed the waste towers of the mill to be emitting large concentrations of hygroscopic salts.

Opposite: A giant thundercloud in which hailstones are carried, rising and falling repeatedly before finally crashing to Earth. Below: Hailstones can grow to huge sizes. These fell on Texas.

thinking big

Plans to alter the weather have not simply been limited to short-term effects, such as releasing rain from a passing cloud or damping down an individual hurricane. One recent scheme involved bringing sunlight to winter. Russian space engineers had long felt that it might be possible to reflect light from the sun onto areas of their country which suffered the long winter darkness of the far northern hemisphere. In 1993 the plan was tested by cosmonauts aboard the Mir space station, who deployed a 80-foot- (25-meter-) wide mirror, made of very fine reflective material weighing only 9 lb (4 kilograms). Turned to catch the dazzling rays of the sun, which in space are unfiltered by any atmosphere, the huge mirror reflected, towards the Earth, a beam of sunlight which spread to 2½ miles (4 kilometers) in width as it touched the surface of the planet. This flash of borrowed sunshine lit up a path through northern Russia for six minutes as Mir sped overhead. In 1999 a second day-long test was planned that would have illuminated several parts of Russia and the Czech Republic, and even flowed on into eastern Germany, but sadly the mirror failed to open properly. Now the experiments will have to await a new champion as Mir has long since been abandoned, and burned up as it fell to Earth.

One of the most bizarre scientific proposals ever contemplated to control the weather was suggested shortly before the First World War by Carroll Livingston Riker, a mechanical engineer who had worked on the construction of the Panama Canal. The plan he proposed to the American president, Woodrow Wilson, in 1913 was breathtaking in its audacity and indeed in its simplicity. Riker proposed nothing less than to redirect the Gulf Stream. Flowing along the coast of Canada

Left: The swirls and eddies of the Labrador current are revealed by the flow of ice in the ocean.

and out to Newfoundland is a stream of cold Arctic water known as the Labrador Current. Off Newfoundland, at the Grand Banks, it meets the warm Gulf Stream current and pours freezing water into the fast-moving warm water from the south, creating dense fog in the region as the warm, moist air over the Gulf Stream is suddenly cooled. The Gulf Stream itself cools as a result of the injection of cold water, and turns south-east. Riker proposed that if the Labrador Current could be blocked, and kept apart from the Gulf Stream, the warm waters would continue on their original north-eastward path and a huge amount of heat energy would be transported closer to Greenland, Britain and northern Europe, each of whose climates would improve dramatically. At the same time, he pointed out, the fogs of the Grand Banks would disappear and icebergs carried into the Atlantic by the Labrador Current would cease to be a menace to shipping.

Being an engineer, Riker planned to use the vast quantities of sand and gravel transported by the Labrador Current to build up a sandbar that would trap the current itself. He suggested placing a 200-mile- (300-kilometer-) long obstacle across the ocean bed. This would cause the sand to pile up, over time reaching the surface, until the Labrador Current was stemmed. It was an astonishing idea and was met with mixed reactions, including worries from fishermen who relied on the rich fish stocks that drew on the nutrients that flowed with the Labrador Current. Riker himself was apprehensive about playing God like this, and proposed a risk-assessment study to begin with, to model the possible consequences. In the end the costs of the First World War intervened and the US Congress did not feel able to allocate the necessary funds for the scheme. The world had a narrow escape. More recent ocean and climate modeling has revealed that the effect of Riker's sandbar would indeed have been a warming of the climate in Europe—and also a catastrophic melting of the Greenland ice-cap and a gradual deluge down the eastern seaboard of the USA.

stopping the conveyor

Carroll Riker's grand plan would probably have had another dramatic effect, one that modern-day climate scientists are only just beginning to understand.

Chapter Three told the story of the thermohaline conveyor (pp.105–7), the global-scale circulation of water through the deepest parts of the ocean, which begins in the North Atlantic with an extraordinary descent of water from the surface to a depth below 3,000 feet (1,000 meters). What drives that descent is the cooling of the salty water carried there by the Gulf Stream. As it cools it becomes more dense, and so it begins to sink. This process is accelerated by the freezing of the surface waters to make sea ice because, as the water freezes to pure ice, the salt is left behind in the liquid sea below. The saltier the water becomes, the more dense it becomes and therefore it descends faster.

But this remarkable circulation is slow and subtle, sensitive to the tiniest changes in the temperature and salinity of the water. The flood of fresh water from the melting ice that would have resulted from Riker's crazy scheme would have significantly reduced the saltiness of the water and it would have failed to sink. If that happened, the driver of the thermohaline conveyor could switch off and the ocean circulation could stall. Scientists have begun to realize that the consequences could have been terrifying. Ocean circulation lies at the heart of the global climate and of the change that is already beginning to overtake us.

a cooler, warmer world

In the last twenty years climate change has been at the forefront of our minds when we think of the weather. What will happen to where we live? How will it be different? As the climate modelers have perfected their science it has become possible to glimpse future states and this has turned up some surprising scenarios. One of the most shocking revelations has been that climate change may not always come gradually, with the world slowly warming over a period of centuries. Instead, the planet may react suddenly and dramatically over just a few short years, and in ways that may seem totally the opposite to what we might expect. These are not just theoretical predictions: it has happened before.

In the mid-1960s a scientist, Russell Coope, found a perfect spot, in the layers of an eroding mud cliff at St. Bees on the north-west coast of England, where he could study dead ground-beetles. Thousands of years ago the carnivorous beetles had happily lived around the edge of a muddy pool, failed to look where they were going and often fallen in and died, so layer upon layer of dead beetles were trapped in the sediment, generation after generation. These beetles are particularly sensitive to temperature—different species will survive only in cooler or warmer climates—so the mud pool provided a remarkable record of the climate for thousands of years. The sediments dated from some 13,000 years ago, the time when the world was emerging from the last Ice Age, and Coope therefore expected the beetle layers to change gradually from cold-loving species at the bottom towards warm-loving above. But to his surprise the climate appeared to have shifted dramatically: warmer, then suddenly cooler, then warming again. About 13,000 years ago the warming world appeared to have sunk back in to another Ice Age that lasted for over a thousand years.

Coope's work remained unnoticed or disbelieved for almost thirty years because the standard view was that the Ice Age had slowly come and gone, with the world gradually warming over many, many centuries as the ice melted—and a few dead beetles weren't going to change that view. But then came the first results of one of the most surprising and important science undertakings to have been carried out within the last generation: the Greenland Ice Core Project (GRIP). It is a massive enterprise that has the simple aim of finding prehistoric snow. By drilling deeper and deeper into the Arctic ice and measuring the composition of the ice crystals (see p.219) scientists during the late 1980s and early 1990s built up a record of prehistoric temperatures far back to the time of the deep Ice Age. The picture that emerged was not one of gradual warming as life on the planet surfaced from the grip of the glaciers, but of sudden shifts in the climate in terrifyingly short periods of time. The planet has swung between periods of relative warmth and icy cold, with average temperatures changing by anything up to 22°F (10°C) in less than ten years. In particular, around 13,000 years ago, as the world was generally warming, there occurred the sudden reversal in climate that Russell Coope had seen in his beetle graveyard—and it took place within just a few decades. This extraordinary time of freezing, in the midst of a warming world, is known as the "Younger Dryas," after a plant that thrives in the tundra, and which flourished in northern Europe for the duration of that icy millennium. But how could the climate change so rapidly; and could it occur again?

It seems that amid the general warming of the planet, and the melting of the great glaciers that had been part of early human life for so long, there was a sudden rush of meltwater from the ice sheets that had lain across North America.

Climate change may not always come gradually ...the planet may react suddenly and dramatically over a few short years ●

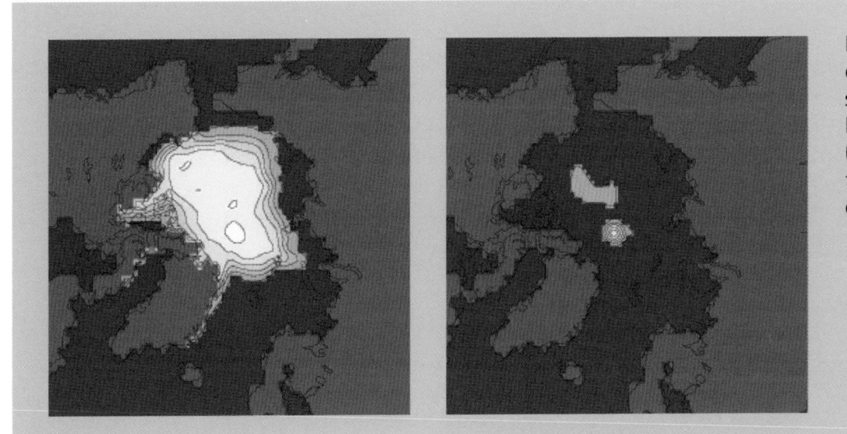

Left: A climate model estimate of the extent of sea-ice over the North Pole at the present day (left) and as predicted at the end of the 21st century (right).

This flowed into the path of the Gulf Stream at exactly around this time and this sudden influx of fresh water would have dramatically changed the saltiness of the sea and had devastating consequences for the thermohaline conveyor. In simple terms, the conveyor shut down and triggered a breakdown of the Gulf Stream, so that warm water no longer flowed north-east across the Atlantic Ocean.

The Gulf Stream is the reason why western Europe is on average much warmer than should be expected for its high latitude on the globe: as the warm water moves north-east and cools it releases heat into the atmosphere, bringing warmer weather. If the conveyor slowed down or shut down, that current flow would weaken and shift south, well away from the North Atlantic, allowing the air above that region to cool. Thirteen thousand years ago the sudden cooling would have begun to build up a cap of sea ice which would have prevented the Gulf Stream starting up again, and the pattern was set for many centuries—a thousand-year setback in the world's recovery from the Ice Age.

Climate scientists are now convinced that the conveyor has switched on and off many times in the past, and global warming today threatens to trigger it again. A warmer world will evaporate more water from the oceans, so rainfall will increase in places such as the North Atlantic where so many of the world's cyclonic storms already form. More rainfall means more fresh water. A warmer world means melting of the Arctic ice—and that, too, means more fresh water. No one knows how much, or how quickly it will appear, but the conveyor is very sensitive to change and the likelihood of it switching off again is no longer remote. If it did so, colder winters would be the immediate result. It has been calculated that Britain would experience an average temperature drop of 3.6°F (2°C) in winter, bringing the average winter temperature down to the level of the coldest year since records began almost 250 years ago. This average would be colder than the "great winter" of 1962–3, (the coldest of the past century) and many even colder winters would be expected—many of them worse than the famous winter of 1683, when the Thames was frozen over with ice a foot (30 centimeters) thick and sea ice blocked the English Channel. This was the time of the Little Ice Age, which lasted for three hundred years from the early sixteenth century. Scientists now think this could well have been caused by an erratic change in the circulation of the conveyor—which may have come very close to switching off altogether during that time.

The consequences for summer would be as extreme, although a little more complex in their cause. The stopping of the Gulf Stream would

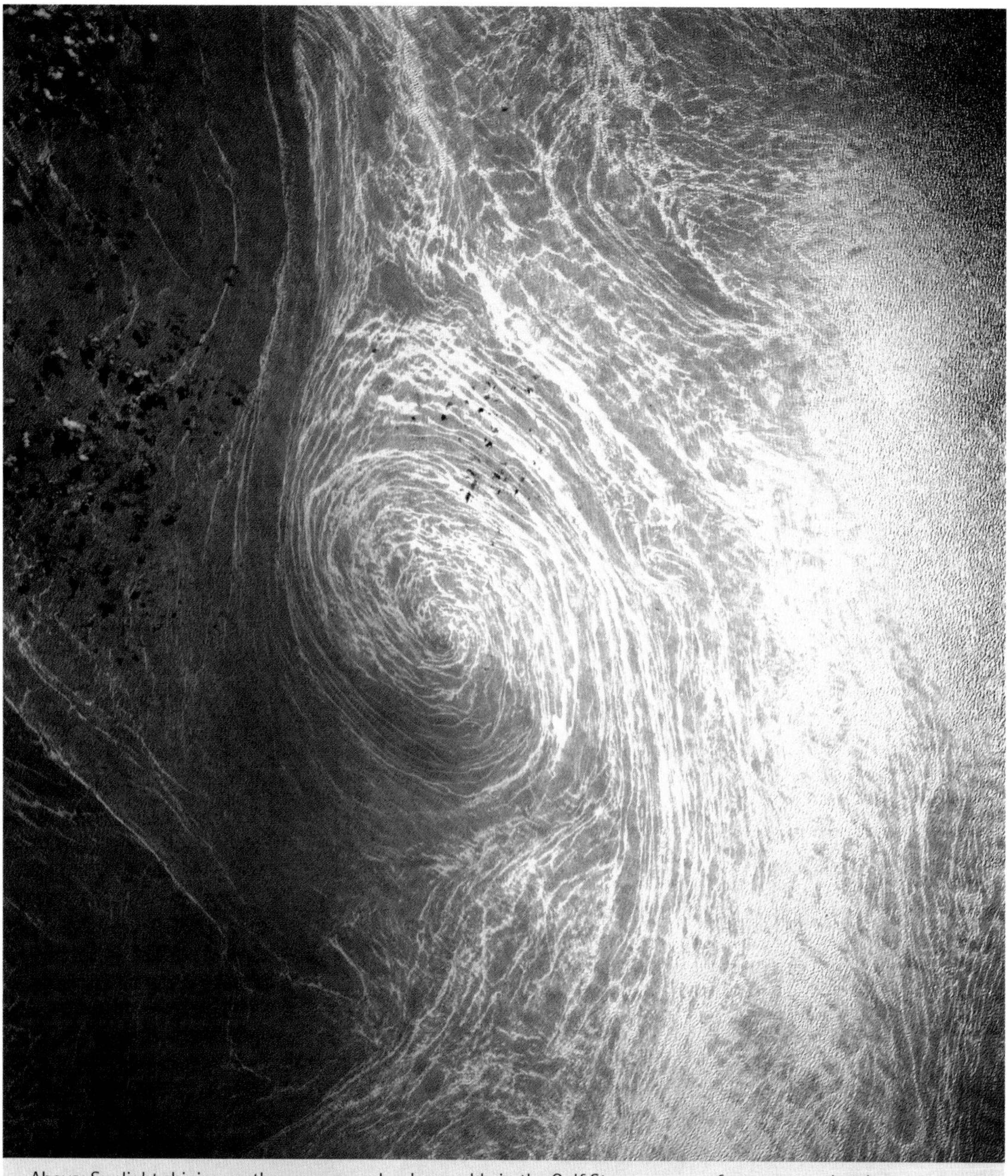

Above: Sunlight shining on the ocean reveals a large eddy in the Gulf Stream, as seen from a space shuttle.

bring an advance of sea ice across the North Atlantic and, together with the lengthier winters and the long-lasting snow cover on land, this vast area of white would reflect back much more of the sun's heat, adding to the cooling effect. So summers would be much shorter and frosts could be expected to continue into June, severely limiting the growing of summer crops. Long term, the extended winters would allow snow cover to build up year after year, reflecting back ever-increasing amounts of sunlight, creating a colder, drier high-pressure zone across much of northern Europe and ensuring that any warmer airflow from the south was kept well away from the cooling Britain. Bizarre as it may seem, global warming could bring us a new Ice Age.

ICE CORES

Snowfall is a measure of the climate, and the properties of every snowflake or snow crystal are fixed by the air and weather through which it falls. The structure of a snowflake is influenced by the temperature at which it falls; and the chemical composition of the ice at its heart reflects the nature of the atmosphere in which it formed—even the time of year when it fell. So if you can find some snow from the distant past you can tell what the weather was like at that time. The problem with snow is that it melts long before any palaeoclimatologist can dig it up—except in two places on the planet: Antarctica and Greenland.

The snow that falls each year on the huge domed center of Greenland's ice cap is gradually covered by layer upon layer of more snow, which slowly compacts to form layer upon layer of ice. But each layer carries with it a telltale sign of its age. The oxygen that makes up air consists of different proportions of oxygen isotopes (atoms with slightly different atomic weights). The vast bulk of natural oxygen is the stable form, O-16, but there is also a tiny trace of the slightly heavier O-18. And it is the O-18 that is the marker, because the proportion of O-18 in the air varies according to the season: there is more of it in summer and less in winter. The scientists realized that if you drill down into the ice and take samples, the peaks and troughs of O-18 will be a mark of each new year. It was a remarkable insight. Drilling down into the 1¾-mile- (3-kilometer-) thick Greenland ice cap, has revealed the composition of the atmosphere, and its temperature, for the last 200,000 years. In Antarctica, at the Russian and French drilling station, the "Vostok" core has gone even further—almost to 500,000 years. The average temperature of the climate for each year is also revealed by the oxygen isotopes, with much more of the heavier one appearing in warmer years. But as well as temperature, stored in the layers are measures of the carbon-dioxide level, the quantities of acid in the atmosphere, the fallout from volcanic eruptions and much more, enabling a remarkably precise portrait of the prehistoric climate to be painted.

Below: Electromechanical ice drills are used by British Antarctic Survey glaciologists to extract ice cores from deep inside the ice sheet. These ice cores are then transported back to England for analysis.

Above: Paintings such as *Hunters In The Snow* by Pieter Brueghel (c.1515–69) have revealed clues to climate change in the recent past.

THE LITTLE ICE AGE

About a thousand years ago the weather across the northern hemisphere was warmer and drier than today: the Vikings successfully colonized Iceland and Greenland, and navigated an ice-free Arctic Ocean to the nearest tip of North America; in England, William the Conqueror's Domesday Book recorded thirty-eight vineyards in operation. But five hundred years later there was a very different picture to be seen. It is hard to know how widespread it was, but from around 1500, Europe was sliding into the Little Ice Age.

In 1982 a British academic, Hubert Lamb, published a book that changed the perception of our past for many climate scientists of the time. Called *Climate History and the Modern World*, it draws together a wealth of evidence from the most extraordinary sources in order to paint a picture of the weather that our forefathers lived through. For Lamb the measurements of the weather could be seen not only in the official temperature records of the last couple of centuries, but also in the weather diaries of a country parson, the fluctuations of wheat and rye prices, the success or failure of wine harvests, the Norse sagas of sea ice coming and going, the poems inspired by dark skies, the paintings of country scenes and even church records of the exorcism of advancing glaciers. All these remarkable insights have enabled a detailed local picture of the climate to emerge. This fleshes out the ever more sophisticated data of global trends that has at the same time emerged from ice cores and ocean-sediment samples. And the period that can be most clearly reconstructed in this way is the Little Ice Age. The Flemish painter Pieter Brueghel was inspired by the harsh winter of 1565 to depict snowy scenes that have become famous today, and in the same year came reports of frost fairs on the River Thames in London. Using a painting of the Rhône glacier in Switzerland in 1750—the depth of the Little Ice Age—can be compared exactly with a photograph taken in 1950 and reveals a huge difference in the advance of the ice three hundred years ago. Lamb even reconstructed weather maps using observations from ships over the two-month lead-up to the Spanish Armada (1588), which revealed fast-moving cyclonic storms driven by a jet stream far stronger than might be expected today—one which would have been caused by a more intensely cold Arctic air-mass. Be it something as tangential as a famine report, or as precise as the poor growth of a forest exposed by the narrowness of the tree-rings, although each individual measurement cannot be accurate on its own, together the picture they build up is consistent. From 1550 to 1800 average temperatures were 0.5°C lower than at the start of the twentieth century. The Little Ice Age held civilization in its grip.

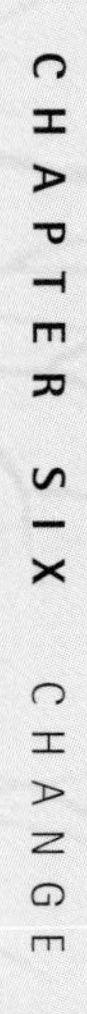

Above: The pollution from an oil-refinery chimney is highlighted by the glow of the sun behind.

the story of global warming

The "greenhouse effect" of the Earth's atmosphere is something that has risen to the top of the public agenda only in the last two decades, but today it is a concept that is spoken of on a daily basis. However, the idea that we could change the climate of our planet has been buzzing around scientific circles for almost two centuries.

In 1827 Baron Jean Baptiste Joseph Fourier, a French mathematician and physicist, spotted that there was a difference between the heat from the sun that was absorbed by the Earth and the heat that was reflected back from the Earth into the atmosphere. He noted that the reflected heat is partly trapped by the atmosphere, which warms up as a result, and he was the first to wonder if human activity could influence the amount of heat the atmosphere retained. Prophetically, he thought that it probably could, although he could not quite see how. In the 1850s an English scientist, John Tyndall, worked out that the two constituents of the atmosphere that retained the heat were water vapor and carbon dioxide, and argued that the relative amounts of these in the atmosphere were the main cause of climate change. But it was a portly, party-loving Swede, Svante Arrhenius, who made the great leap for climate science by calculating the precise effect carbon dioxide could have on the climate. Arrhenius was part of the same circle as Vilhelm Bjerknes, the Norwegian who became known as the "father" of meteorology (see Chapter Three). Weighing almost 220 lbs (100 kilograms), he was a huge man with a huge idea: he believed that the Ice Ages had been caused by vast fluctuations in the volume of carbon dioxide in the atmosphere. Fixed on this idea, Arrhenius set about calculating the amount of change that would be required. It was a huge task. Today even the world's most powerful computers can take a week or more to run a sophisticated model of the global climate that will pick out the planet-wide temperature effects of subtle changes in CO_2 levels. Arrhenius had to do the fiendishly complex calculations by hand. He worked out the warming effect of five different levels of CO_2, at

GLOBAL WARMING

Scientists are still very much debating the issue of global warming—its causes and effects. The global climate is seeing an increasing rise in temperature and it is estimated that by the year 2100 the surface air might have warmed by between 2 and 6°F (1 and 3.5°C).

This rise in temperature is due to the increasing concentration of greenhouse gases in the atmosphere. These gases (most notably carbon dioxide) trap some of the energy coming from the sun and prevent the Earth's outgoing radiation from being lost into space. This "greenhouse effect" has been occurring for millions of years and without it the Earth would be 10 to 15 degrees cooler. However, some human activities added to the amount of greenhouse gas in the atmosphere, and this is now trapping more of the sun's energy and so warming the planet. These human activities include burning fossil fuels, such as coal, which releases more greenhouse gases into the atmosphere, and the destruction of the rainforests, whose trees would normally absorb carbon dioxide. The increased concentration of carbon dioxide is only one cause for concern, as the atmosphere is also seeing increasing amounts of methane, nitrous oxide and CFCs (chlorofluorocarbons).

Below: The global climate of the Earth is starting to see a rise in temperature. This is as a result of increased greenhouse gases in the atmosphere trapping more of the sun's energy.

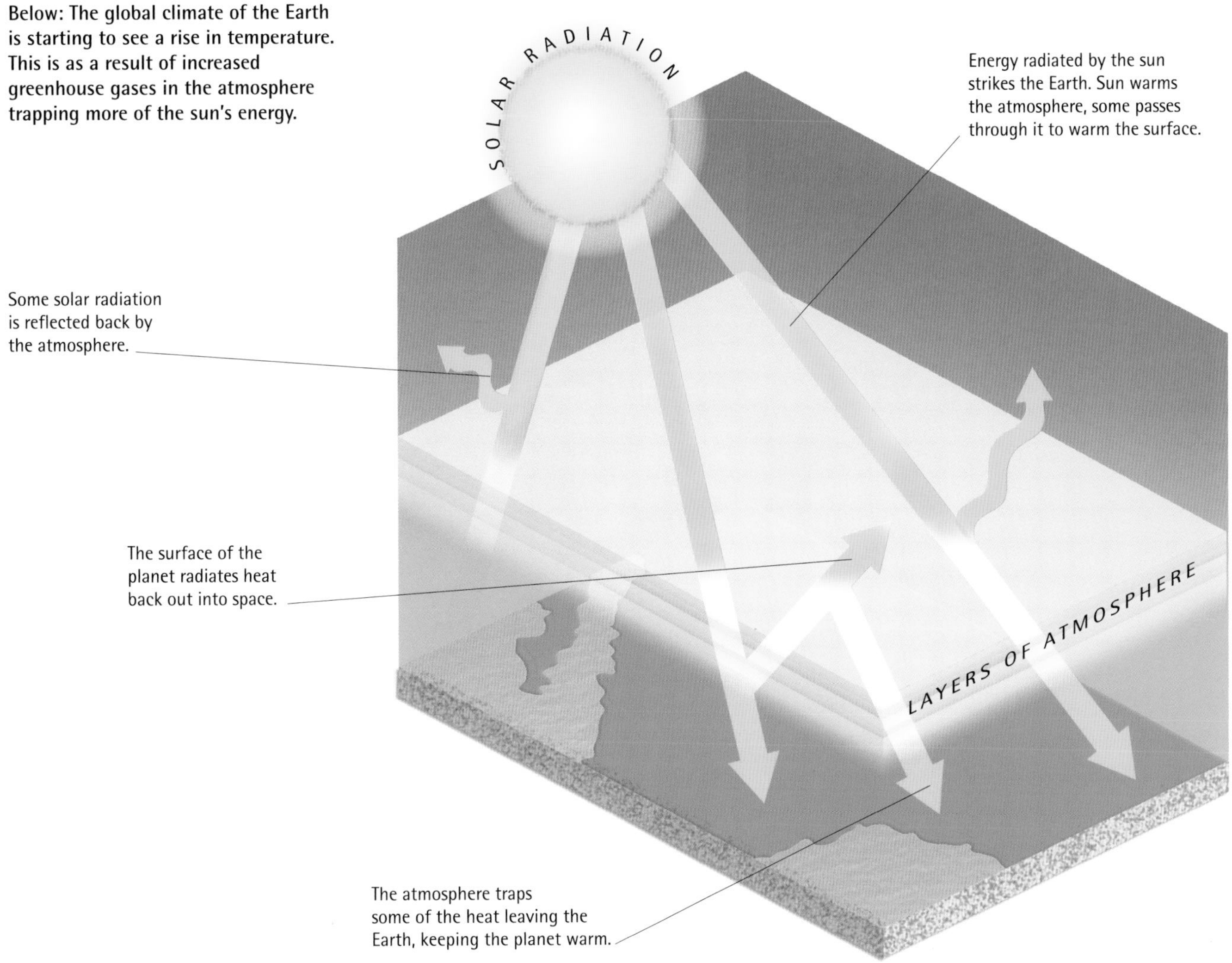

every 10 degrees of latitude, north and south on the globe. Working obsessively, often fourteen hours a day, he spent over a year on the task and arrived at a surprisingly accurate set of tables: a kind of ready-reckoner for climate change. He concluded that the Ice Ages could indeed have been caused by huge reductions in atmospheric carbon dioxide (although we now know that the causes were far more complex); and he predicted that a doubling of the amount of the gas would produce a 9°F (5°C) increase in global temperature (today's estimates are slightly less than this). But his great contribution was to make the link between the burning of fossil fuels by modern industrial societies and a warming effect on the climate. Again, he laboriously calculated how long it would take for the burning of coal to double the carbon-dioxide level. Based on the industrial activity of the late 1890s, his answer came out at about three thousand years. If you feel that the world should have taken heed of this warning, perhaps Arrhenius's own attitude to such global change may provide a touch of irony; he foresaw an era when our distant descendants would bask in a warmer and less harsh climate than that of his fellow compatriots in Scandinavia.

Little attention was paid to the idea that an atmospheric gas could have such a profound effect. Most scientists believed that the oceans would simply absorb any excess carbon dioxide in the atmosphere, holding it for millennia, and so effectively reduce the effect to zero. But then, in the late 1950s, the burgeoning science of oceanography revealed that the oceans do not absorb the CO_2 anything like quickly enough to make a difference, and the realization that humans were tampering with nature on a global scale finally dawned. Since 1958, at the observatory that lies at the top of Mauna Loa volcano in Hawaii, a set of measurements of atmospheric carbon dioxide has been taken regularly, and it is this constant window on the sky that has shown overwhelmingly the effect that we have had on our planet. Carbon dioxide levels in the atmosphere have risen by almost 15 per cent since the first measurements were taken forty years ago, and scientists have calculated that they have increased by over 30 per cent since the start of the Industrial Revolution. This concentration of CO_2 is higher than at any time in the Earth's history. Today, however, it is not just the effect of carbon dioxide that is at work. Since Arrhenius happily predicted a better, warmer world, the arrival of the motor car and industrialization on a scale unimaginable eighty years ago have together generated other potent greenhouse gases, such as nitrous oxide and methane, as well as tapping vastly increased amounts of fossil fuel to burn into the atmosphere. And the result? All in all, there has been an increase in the mean atmospheric temperature near the planet's surface of 1.1°F (0.6°C) since the start of the twentieth century.

Above: A scientist working on air-sampling equipment in front of the buildings of the Climate Monitoring and Diagnostic Laboratory on Mauna Loa, Hawaii.

Left: Global warming will put more energy into the weather system and bring extreme weather to some surprising places. A computer-generated image of a tornado over the White House in Washington.

Above: Rapid cliff erosion in Norfolk, England, provides a taste of the coastal changes that will occur with rising sea levels.

future weather

The latest report of the Intergovernmental Panel on Climate Change has unequivocally pointed the finger of blame for the warming world at our modern human society. So what exactly are we bringing on ourselves for the future? At first the scenario seemed simple. Warmer, drier summers and winters; more drought—a grim prospect. But as time has gone on and the science has become more subtle, the pattern has grown more complex. What global warming will do is add much more energy to the climate system—more heat. Over the great continental interiors of Australia, Africa, Asia and America that may indeed produce drier land. But over the oceans the heat will mean more evaporation of the water, more moist air rising faster in the sky. Quite simply, this will bring more rain, greater differences in air temperature and pressure, and thus faster winds; in short, more violent, wetter storms. Hurricanes are expected to intensify by some 10 per cent over the next century. The El Niño events that engulf the coast of South America will also intensify. Ocean waves are even now getting taller and stronger. We have already seen how the melting Arctic and Antarctic ice could change the nature of the ocean circulation, as more fresh water dilutes the saltiness of the sea, as well as bringing the gradually rising sea levels that are already beginning to threaten the small island states of the Pacific and Indian oceans. So when the increase in rainstorms has faded into familiarity, severe winters may well be in store for the countries of Europe that rely on the warmth of the northern ocean currents.

Will it happen? The signs are that it has already begun. Finding the fabled North-west Passage from the Pacific to the Atlantic Ocean was the dream of explorers in the centuries after the first circumnavigators rounded the globe. A passage through the Arctic Ocean would have meant a clear trade route to and from the spice islands of the Pacific. It might have saved much time and many lives that were otherwise lost on the long ocean voyages around the southern capes. But the passage was never found—always, the solid pack-ice barred the way of intrepid sailors, many of whom became entombed in ice that closed around their vessels. In 1940 a ship of the Royal Canadian Mounted Police finally made the breakthrough. It carved its way through the Arctic ice, taking over two years to complete the journey from west to east, over the top of the North American continent. It was a heroic achievement for a tough ice-breaking ship, and a brave crew. Sixty years later, at the start of the new millennium, the journey was repeated. This time it took little more than a month, and at no stage did the crew encounter pack ice—just isolated floes and occasional icebergs. The difference is startling. Measurements over the last half-century reveal that the Arctic ice is only 60 per cent as thick as it was fifty years ago, and in 2000 clear blue water instead of ice was seen over the North Pole itself.

The signs of change are no longer buried in scientific data. They are here for us all to see.

Above: Kiribati—this tiny, low-lying island in the South Pacific is typical of the lands that some scientists believe may disappear with rising sea level.

bibliography

William K Stevens, *The Change in the Weather,* (Delacorte Press, 1999)

Alan Villiers, *Wild Ocean* (McGraw-Hill, New York, 1957)

Louise B. Young, *Earth's Aura* (Allen Lane/Penguin Books, 1977)

A. B. C. Whipple, *Planet Earth: Storm* (Time-Life Books, 1982)

Willis H. Miller, 'Santa Ana Winds and Crime', *Professional Geographer*, Vol XX, No 1, Jan 1968

Lyall Watson, *Heaven's Breath* (Hodder & Stoughton, 1984)

F. G. Sulman et al, 'Absence of Harmful Effects of Negative Air Ionisation', *Int J Biometeor*, Vol 22, no 1, 1978

Thomas P. Grazulis, *Tornadoes of the United States,* (University of Oklahoma Press)

Robert Henson, 'Billion Dollar Twister', *Scientific American*, Spring 2000

Paul Simons, *Weird Weather,* (Little Brown & co, 1996)

John Rousmaniere, *Fastnet Force 10* (Norton, 1993)

Christopher Torrence and Peter J Webster, *Jo Climate, 12*

Paul J Kocin et al, 'Overview of the 12-14 March 1993 Superstorm', *Bulletin American Meteorological Society*, Vol 76, No 2, Feb 1995

acknowledgements

In the process of writing this book, I have been inspired and enthused by the work of many authorities on meteorology, climate and survival, whose far more expert publications make enthralling further and deeper reading. I would like to express my gratitude to them all. And my thanks go to Stan Cornford who took such trouble to read my words and identify numerous errors—although I alone retain responsibility for any that remain.

I would also like to thank the production team of BBC Science's *Wild Weather* series for creating the fine programs without which the book would be infinitely the poorer. I am particularly grateful to Will Aslett, the series producer, who crafted the shape of the series, which these chapters mirror closely, and to Jill Rankin who undertook the very wide-ranging and meticulous initial research from which both book and television programs were built. And throughout the process my appreciation has gone also to my editor, Helena Caldon, for her calming advice, to Sarah Ponder, and to my assistant, Tara Prayag, who has always been there to help me manage my deadlines.

But above all I owe a debt of gratitude to my wife, Ewa Hawrylowicz, and my sons Christopher and Toby, for their months of patience, their cheerful encouragement, and their acceptance of lost weekends and holiday time.

picture credits

Adrian Meredith Aviation Library/British Airways Archive Photographs page 55; Arctic Institute Denmark 138; Brian and Cherry Alexander 24, 124-25, 133, 134, 135, 160, 202-203; The Art Archive 23; The Bridgeman Art Library: 24, 34-35, 38, 39, 136-137, 220; The British Library 99; Bruce Coleman 175, 191; Corbis 42-43, 58-59, 87 (inset), 218, 222, 227; Cornelius Braun 147; Ecoscene/Jim Winkley 198; Frank Lane Picture Agency 19, 36, 212; Genesis Photo Library 119 (inset); Hulton Getty Images 26, 122, 139, 156, 159(inset); Hutchison/Nick Haslam 82, 83; Image Bank 30, 157, 170; Images Colour Library 195 (left); Jonathan Renouf 54; Knudsen Fotosenter 80,130; Image Bank/Carolyn Brown 92; Landform Slides 182; MeteoSwiss: 60-61 (©Beat Kaslin); NASA/Science Photo Library 57; National Archives of Canada 53 (inset); National Met Library © M Samson 151 (inset); Oxford Scientific Films 33 (micro), 47 (below right), 48-49, 62-63, 64, 66, 70-71 (background), 80-81, 84-85, 88-89, 97, 104-105, 140, 142, 158 (bg), 176, 185 (1&3), 188-189, 190-191, 196, 213; PA Photos/EPA 100-101, 110, 164-165; Photo Researchers Inc/Vanessa Vick 154; George Plafker/US Geological Survey 160; PPL Ltd 72; Robert Harding 123; Royal Astronomical Society Library 41; Royal Meteorological Society 98; Science Photo Library 10-11, 11 (micro), 14-15 (inset), 15, 16-17, 18, 20 (above and below), 21 (above and below), 27, 28 (inset), 31, 32-33, 43 (inset), 47 (below left), 50, 74, 75, 77 (inset), 78-79 (both),112, 116,117, 118, 102-3, 128-129, 132 (inset), 143, 146, 148, 149, 162-163, 177, 184, 185 (2&4), 187, 195 (right), 208, 210, 219, 224-5; South American Pictures 172-173; Still Pictures 183, 200-201; Tony Stone Images 12-13, 166-7, 168, 169, 180-181; Topham Picturepoint 45, 73, 95, 120; Woodfall Wild Images 87, 150-151, 226; © Paul Welti 116 (right).

index

Entries in *italic* refer to illustrations

d

e

f

g

h

i

j

k

l

m

A FIREFLY BOOK

Published by Firefly Books Ltd., 2002

First Printing

National Library of Canada Cataloguing in Publication Data

Lynch, John
The weather / John Lynch.

Includes bibliographical references and index.
ISBN 1-55297-640-8 (bound).–ISBN 1-55297-639-4 (pbk.)

1. Meteorology. 2. Weather. I. Title.

QC981.2.L95 2002 551.5 C2002-902223-1

Publisher Cataloging-in-Publication Data (U.S.)

Lynch, John.
The weather / John Lynch.–1st ed.
[240] p. : col. ill., maps ; cm.
Includes bibliographical references and index.
Summary: An illustrated guide to weather terminology, phenomena and weather pioneers.
Companion book to the BBC/The Learning Channel Series.
ISBN 1-55297-640-8
ISBN 1-55297-639-4 (pbk.)
1. Weather. 2. Meteorology. 3. Weather forecasting. I. Title.
551.6 21 CIP QC861.4.L96 2002

Published in Canada in 2002 by
Firefly Books Ltd.
3680 Victoria Park Avenue
Toronto, Ontario M2H 3K1

Published in the United States in 2002 by
Firefly Books (U.S.) Inc.
P.O. Box 1338, Ellicott Station
Buffalo, New York 14205

This book is published to accompany the television series entitled *Wild Weather*, which was produced by the BBC and first broadcast on BBC1 in 2002.

Executive Producer: Bill Grist
Series Producer: Will Aslett
Producers: John Maguire, Ben Fox

Published by BBC Worldwide Ltd.,Woodlands, 80 Wood Lane, London W12 0TT

Commissioning Editor: Sally Porter
Project Editor: Helena Caldon
Text Editor: Ruth Baldwin
Cover Art Direction: Pene Parker
Book Art Direction: Sarah Ponder
Book Design: Paul Welti
Academic Consultant: Stan Cornford, M.Sc.
Production Controller: Kenneth Mckay
Picture Research: Frances Abraham, Rachel Jordan, Miriam Hyman, Claire Parker
Illustrations: Andrew O'Brien, Olive Pearson

Set in Rotis Semi Sans Light and Cosmos
Jacket printed by Lawrence Allen Ltd, Weston-super-Mare
Color reproductions by Kestrel Digital Colour, Chelmsford
Printed and bound in Great Britain by Butler & Tanner, Frome & London